DECARBONISING ELECTRICITY

The current shift to renewable energy is dominated by globalised energy companies building large-scale wind and solar plants. This book discusses the consequences and possibilities of this shift in India, Germany, and Australia, focusing on regions that have now largely decarbonised electricity generation. The authors show how centralised models of energy provision are maintained and chart their impacts in terms of energy geography, social stratification, and socio-ecological appropriation. The chapters emphasise the prominent role played by state regulation, financial incentives, and public infrastructure for corporate renewables, arguing that public provision should be repurposed for distributed renewables, social equity in affected regions, and wider social benefit. This interdisciplinary book provides fertile building ground for research in – and application of – future energy transitions. It will appeal to students, researchers, and policymakers from anthropology, sociology, politics and political economy, geography, and environmental and sustainability studies.

DECARBONISING ELECTRICITY

The Promise of Renewable Energy Regions

JAMES GOODMAN

GARETH BRYANT

LINDA CONNOR

DEVLEENA GHOSH

JONATHAN PAUL MARSHALL

TOM MORTON

KATJA MÜLLER

STUART ROSEWARNE

RIIKKA HEIKKINEN

LISA LUMSDEN

MAREIKE PAMPUS

PRIYA PILLAI

CAMBRIDGE
UNIVERSITY PRESS

CAMBRIDGE
UNIVERSITY PRESS

Shaftesbury Road, Cambridge CB2 8EA, United Kingdom

One Liberty Plaza, 20th Floor, New York, NY 10006, USA

477 Williamstown Road, Port Melbourne, VIC 3207, Australia

314–321, 3rd Floor, Plot 3, Splendor Forum, Jasola District Centre, New Delhi – 110025, India

103 Penang Road, #05–06/07, Visioncrest Commercial, Singapore 238467

Cambridge University Press is part of Cambridge University Press & Assessment, a department of the University of Cambridge.

We share the University's mission to contribute to society through the pursuit of education, learning and research at the highest international levels of excellence.

www.cambridge.org
Information on this title: www.cambridge.org/9781009485616

DOI: 10.1017/9781009485609

When citing this work, please include a reference to the DOI 10.1017/9781009485609

First published 2025

A catalogue record for this publication is available from the British Library

A Cataloging-in-Publication data record for this book is available from the Library of Congress

ISBN 978-1-009-48561-6 Hardback

Contents

Contributors

Gareth Bryant is Australian Research Council DECRA Fellow and Chair of the Discipline of Political Economy at the University of Sydney. His research explores the political economy of sustainability and inequality in a range of policy areas, including climate change, energy, housing, and education. He is author of Carbon Markets in a Climate-Changing Capitalism (Cambridge University Press, 2019).

Linda Connor is Emeritus Professor of Anthropology at the University of Sydney. She has conducted long-term ethnographic research on coal mining conflicts, climate change, and renewable energy in Hunter Valley communities and in South Australia. Her publications include Climate Change and Anthropos (Routledge, 2016).

Devleena Ghosh is Honorary Professor in Social and Political Sciences at the University of Technology Sydney. She has published widely in colonial, global, and environmental history and is the co-author of Colonialism and Modernity (UNSW Press, 2007). She is currently working on a project on indigenous responses to fossil fuel extraction in India.

James Goodman is a Professor in Social and Political Sciences at the University of Technology Sydney. He has led two large Australian Research Council projects: Beyond the Coal Rush (2014–2017) and Decarbonising Electricity (2018–2024). He is a co-author of Justice Globalism (Sage, 2013) and Climate Upsurge (Routledge, 2014).

Riikka Heikkinen works in the renewable energy industry and is a PhD candidate with the University of Technology Sydney. She is researching large-scale renewable energy support mechanisms and their impacts and is affiliated with several renewable energy research projects. She works hands-on with communities on large energy projects.

Lisa Lumsden is a regional energy transition practitioner with The Next Economy. She has lived experience of energy transition, having served as Port Augusta City Councillor during the surprise closure of the Playford and Northern coal power stations in South Australia and the subsequent rise of renewables. She is a recipient of the Jill Hudson Award for Environmental Protection.

Jonathan Paul Marshall is a Senior Lecturer at the University of Technology Sydney, researching the problems of energy transition. He is co-editor of Environmental Change and the World's Futures (Routledge, 2016) and has also edited several journal issues on climate and energy, including Energy Policy and Energy Research and Social Science.

Tom Morton is an associate member of the Centre for Interdisciplinary Regional Studies (ZIRS) at the Martin Luther University Halle-Wittenberg and is based in Sydney. He is co-editor of From Carbon Democracy to Climate Democracy (Routledge, 2024) and works as a freelance journalist and media convenor for the climate action group Rising Tide in Australia.

Katja Müller works in the fields of energy humanities, as well as digitisation, museum studies, material culture, and visual anthropology. With a background in anthropology, she is Heisenberg-Professor for Technology, Ethics and Society at Merseburg University for Applied Sciences, conducting research into digitisation processes in energy transitions.

Mareike Pampus is a postdoctoral researcher and lecturer in the Human Geography Department at Martin Luther University Halle-Wittenberg and also serves as a scientific coordinator at the Centre for Just Transition and Sustainability. Her current postdoctoral research explores diverse nature concepts in land restoration of former lignite mining areas in the Central German Mining District.

Priya Pillai has worked on environmental and social justice issues for over 20 years, including with Greenpeace, the Right to Food Campaign, Action Aid, and Oxfam, and is associated with the National Alliance of People's Movement (NAPM). She is a PhD candidate at the University of Technology Sydney, researching the socio-ecological impacts of large-scale renewable energy in India.

Stuart Rosewarne is an Honorary Associate Professor in Political Economy at the University of Sydney, focusing on environmental and ecological economics, socialist ecology, international political economy, and the political economy of gender. He is the author of Contested Energy Futures: Capturing the Renewable Energy Surge in Australia (Palgrave, 2023).

Acknowledgements

We would like to thank all the research participants who shared their time with us for this book. We are very grateful for their many insights into the impacts of large-scale renewable energy. We also would like to thank the additional researchers, interpreters, transcribers, photographers, and journalists, as well as the administrative workers at our institutions who helped us to establish and complete the research project on which this book is based. Also, we would like to thank our respective universities and other institutions that have supported us in undertaking the project. Thanks are also due to the Australian Research Council, which supported the project between 2018 and 2023 through a Discovery Project, 'Decarbonising Electricity: A Comparison in Socio-ecological Relations'. This book would not have been possible without the commitment of this Australian governmental agency to explore key challenges that face societies, not least among them the challenge of climate change and decarbonisation. Thanks are also due to Partner Investigators in the Project, Professor Dipesh Chakrabarty at the University of Chicago and Professor Ortwin Renn at the Research Institute for Sustainability – Helmholtz Center Potsdam. Finally, we would like to thank Cambridge University Press and their staff: it has been a great pleasure to work with them to bring this book to fruition.

Introduction

Legitimacy for Renewables?

A Prologue from the Future

> 'It's amazing when you think of it', said Adell. […] 'All the energy we can possibly ever use for free. Enough energy, if we wanted to draw on it, to melt all Earth into a big drop of impure liquid iron, and still never miss the energy so used. All the energy we could ever use, forever and forever and forever'.
>
> —*Isaac Asimov,* The Last Question *(1956: 8).*

> There is no time for a new energy system to evolve gradually over centuries, as was the case for the fossil fuel-based system. […] The energy transition must … become a strategic tool to foster a more equitable and inclusive world.
>
> —*Francesco La Camera Director-General, International Renewable Energy Agency (2023: 5)*

It may seem bizarre, perhaps even perverse, while facing the hottest years on record, with UN Secretary-General António Guterres announcing the era of 'global boiling', to begin a book on one of the most urgent dilemmas of our time with a quotation from a short science-fiction story written nearly 70 years ago (Guterres 2023; Niranjan 2023). But Isaac Asimov's story *The Last Question* neatly encapsulates some of the utopian fantasies associated with the promise of renewable energy. The story begins in 2061, a year now not all that distant, and only a decade beyond the mid-century deadline which the Paris Agreement sets for the global economy to achieve net zero emissions. Two technicians, Adell and Lupov, the 'faithful attendants' of the giant supercomputer Multivac, sit drinking in a deserted underground chamber ruminating on Multivac's latest achievement (Asimov 1956). Faced with the ever-increasing demand for energy to sustain human life on Earth and power interplanetary exploration, Multivac has designed and built a new and apparently inexhaustible energy source:

> The energy of the sun was stored, converted, and utilized directly on a planet-wide scale. All Earth turned off its burning coal, its fissioning uranium, and flipped the switch that connected all of it to a small station, one mile in diameter, circling the Earth at half the distance of the Moon. All Earth ran by invisible beams of sunpower (Asimov 1956: 7).

While Adell celebrates Multivac's technological triumph – 'all the energy we can possibly ever use for free … forever and forever and forever' – his colleague Lupov sounds a warning note. 'Not forever', objects Lupov, pointing out that the 'invisible beams of sunpower' will only last as long as the lifespan of the Sun itself – around 10 billion years. Adell observes that this is probably long enough ('it will last our time, won't it?'), but as the story progresses through 'several trillion years of human history', we learn that Multivac's solution has only proved temporary. Once interstellar travel became possible, what had seemed an infinite supply of energy is nowhere near enough as humans colonise space: 'It took mankind a million years to fill one small world and then only fifteen thousand years to fill the rest of the Galaxy'. Human and non-human ingenuity and technological innovation run up against a fundamental limit: 'the net amount of entropy of the universe' (Asimov 1956: 9).

Present-Day Crisis

The fundamental limit made visible by the climate emergency is much closer to home, arising from the planetary boundaries of the biosphere, which sustains human life on Earth. These boundaries have been destabilised primarily by the burning of coal, oil, and gas, and by the hegemony of 'fossil capital' over the last 250 years. But as Francesco La Camera, Director-General of the International Renewable Energy Agency (IRENA), notes in his introduction to the *World Energy Transitions Outlook 2023*, 'there is no time for a new energy system to evolve gradually over centuries, as was the case for the fossil fuel-based' system (IRENA 2023: 13). According to the Sixth Assessment Report of the Intergovernmental Panel on Climate Change (IPCC), the total remaining carbon budget, if global heating is to be kept below 1.5°C, is no more than 400 gigatonnes (Gt) (IPCC 2021). At present rates of annual global emissions, we will exhaust that carbon budget in less than 7 years (MRIGCCC 2021). The budget was later revised downwards, to 250 $GtCO_2$ as of January 2023, equal to around six years of current CO_2 emissions (Lamboll et al. 2023).

The IPCC, the International Energy Agency, and the Director-General of the United Nations (though apparently not the current President of the COP) are unanimous in their assessment: only a worldwide moratorium on new fossil fuel projects and a rapid and comprehensive transitioning of the world's energy system to renewable sources can avert a catastrophic breaching of the planetary boundaries of the Earth's climate system. According to the IEA, global renewable energy capacity must grow threefold by 2030 in order for the world to remain on a pathway to 1.5°C (IEA 2024).

Yet despite the urgency of this transition, and the need for what IRENA calls 'profound and systemic transformation of the global energy system', there is resistance, not only from the fossil-fuel industrial complex and its supporters in the media, institutions, and both mainstream and populist right-wing political parties, but on the ground, in the villages, places, and spaces where people encounter the expansion of renewable energy in their daily lives. Instead of embracing the promise of renewable energy, some reject it, or regard it with cynicism and mistrust. This book sets out to understand the reasons for this resistance, and the forms it takes, drawing on ethnographic case studies conducted in India, Germany, and Australia. While much of the local opposition to wind and solar farms appears to grow out of purely local concerns – who is benefiting and who is not, a lack of consultation and participation, the physical and visual impacts on local landscapes and their uses for agriculture and leisure – we argue that the deeper reasons must also be sought in the distinctive trajectory of the energy transitions we have studied, one shaped by what we call the neoliberalisation of renewable energy. In order to fully realise the promise of renewable energy, we contend, we must move beyond the neoliberal model of transition to a re-commoning of energy, one that no longer serves the pursuit of continuous economic growth.

Private Renewables – Models and Limits

The neoliberal model of transition, whose characteristics we document in the first section of the book, collides with boundaries which are not primarily the biophysical boundaries of planetary climate stability, but rather are social, political, and economic in nature. One of these is captured in an earlier report from IRENA, the *2021 World Energy Transitions Outlook: 1.5°C Pathway*. The report points to an immanent potential of renewable energy: the 'almost unlimited compression of clearing prices' (IRENA 2021: 163). In contrast with non-renewable forms of energy such as fossil fuels, renewable energy, like the 'invisible beams of sunpower' in Asimov's story, is at least potentially inexhaustible. Once the initial monetary and energy costs of building renewable energy are covered, additional electricity generation is effectively free, because unlike fossil fuels, there are no continuing fuel costs for solar and wind, only the costs of maintenance. Arguably, for the first time in human history (or at least since the publication of Asimov's story), we are presented with the possibility of nearly free universal and inexhaustible energy. As such, renewable energy holds out the promise of energy becoming a commons, but for the authors of the 2021 IRENA report, this is a problem, rather than something to be celebrated.

The problem consists in the lack of incentives for investors. IRENA's modelling points to the emergence of zero marginal cost for daytime electricity. But IRENA's blueprint for delivering the energy transition, with the urgency which climate science dictates, relies on the private sector to deliver 90% of investment finance. As renewables become cheaper, however, renewable energy ceases to deliver an 'acceptable' return on capital in liberalised energy markets designed according to principles of marginal pricing. As IRENA puts it, 'the more renewable energy enters the system, the lower its remuneration becomes, reducing prospects for cost recovery and paralysing new investments' (IRENA 2021: 163). Over time, with technology leapfrogging the necessity for large-scale renewable utilities, this public–private model is unsustainable. IRENA acknowledges that the 'misalignment' between the uptake of renewables and the structure of liberalised energy markets will necessitate a 'comprehensive rethinking of power system structures' (IRENA 2021: 164).

The boundary that becomes visible here is thus a boundary created by fossil capital itself. In the fossil fuel era, as William Stanley Jevons famously observed, demand for coal actually increased, rather than declined, as technological innovation made its use as an energy source more efficient. Asimov's story echoes the Jevons paradox in a science-fiction setting; 'all the energy we can possibly ever use for free', supplied by one small station orbiting the Earth, does not lead to a stabilisation of humans' energy use, but to an expansion, as ever-more energy is required to power the colonisation of other worlds. In one sense, we are still trapped in the Jevons paradox today; global fossil fuel use continues to expand along with the rapid growth of renewable energy. Instead of renewable energy replacing coal, oil, and gas, it appears to be simply complementing them as global energy demand grows. We assume that a plentiful supply of cheap energy will underpin a 'burgeoning global expectation of continuous economic growth, material accumulation and "progress"' (Strauss et al. 2013: 11).

According to the most recent analysis by the International Energy Agency, however, the peak moment for fossil fuels may be approaching: 'the momentum behind clean energy transitions is now sufficient for global demand for coal, oil and natural gas to all reach a high point before 2030' (IEA 2023: 18). Under the IEA's Stated Policies Scenario, 'the share of coal, oil and natural gas in global energy supply – stuck for decades around 80% – starts to edge downwards and reaches 73% by 2030' (IEA 2023: 18). Surveying the latest trends in the development of renewable energy, the IEA concludes that 'a pathway to limiting global warming to 1.5°C is very difficult – but remains open' (IEA 2023: 17).

More recent analysis from IRENA complements that of the International Energy Agency and offers a policy blueprint for how a 'pathway to 1.5°C' might be achieved. IRENA's *World Energy Transitions Outlook 2023* begins the 'comprehensive rethinking of power system structures' which the earlier *Energy Transitions Outlook 2021* envisaged (IRENA 2021: 164). The language of these reports may appear abstract and bloodless, too far removed from the brutal realities of the floods, fires, famines, droughts, and extreme temperatures which have already killed tens of thousands of people, displaced hundreds of thousands more, killed millions of non-humans, and destroyed their habitats (Dunne 2023; NOAA 2023; UNDRR 2023). But they signal an emerging paradigm shift in the framing of climate and energy policy by influential international institutions: IRENA explicitly states that the energy transition must become 'a strategic tool to foster a more equitable and inclusive world', and calls for 'systemic transformation', a recognition that business as usual will not deliver the pathway to 1.5°C the world needs (IRENA 2023: 5).

The *World Energy Transitions Outlook 2023* compares two scenarios: the 1.5°C Scenario, an 'energy transition pathway aligned with the goal … to limit global average temperature increase by the end of the present century to 1.5°C', and the Planned Energy Scenario, which 'is based on governments' energy plans and other planned targets and policies in place at the time of analysis' (IRENA 2023: 17). Broadly speaking, the 1.5°C Scenario describes what needs to happen to limit global temperature rise to 1.5°C; the Planned Energy Scenario describes what is actually happening. The 1.5°C Scenario requires cutting CO_2 emissions by around 37 Gt from 2022 levels and achieving net-zero emissions in the energy sector by 2050.

In order for this to happen, to give just one example, the global percentage of electricity generated from renewables must rise from the current figure of 28–68% by 2030 and 91% by 2050 (and the global percentage of electrical energy use in the total energy supply must also increase massively). In what may amount to the understatement of the century, IRENA observes that 'the energy transition is off-track' (IRENA 2023: 21). The gap between the 1.5°C Scenario and the Planned Energy Scenario, between what needs to happen and what the signatories to the Paris Agreement are actually doing, could best be described as a yawning chasm.

The 1.5°C pathway requires 1,000 GW of renewable power to be deployed every year from now until 2050, but in 2022, only 300 GW of renewables were added to global generation capacity. The share of renewable energy in the global energy mix must increase from 16% in 2020 to 77% by 2050 in IRENA's 1.5°C Scenario. This is the share of renewable energy in total primary energy supply, not just electricity generation; according to the report, 'total primary energy supply would remain stable due to increased energy efficiency and growth of renewables' (IRENA 2023: 25). In order to achieve these goals, what IRENA describes as 'an enduring investment gap' must be overcome; annual investment of US$5 trillion is required; despite global investment in 'all energy transition technologies' reaching a record level of US$1.3 trillion in 2022, the figure must more than quadruple to remain on the 1.5°C pathway (IRENA 2023: 25).

Overall, the emissions reduction goal is to be achieved only partly by renewables and electrification (respectively, 25 and 19%); it will also rely on energy conservation and efficiency (25%), along with a combination of hydrogen, carbon storage, biofuels, and nature-based offsetting (31%) (IRENA 2023: 52). Efficiency is central: IRENA states that in 2050 total global energy consumption will need to be about 5% below 2020 levels (IRENA 2023: 48). Averaged over the period from 2023 to 2050, economic growth is expected to remain at 1.5% annually, which equates to a more than 50% increase in the size of the world economy. Business-as-usual does deliver some efficiency gains, for instance, with IEA projections suggesting that with existing policies energy consumption will grow by 25% over the period. This underlines the efficiency gap to be filled: the ambition is heroic, and at odds with experience in which energy consumption rises with rising income and falling energy prices (Diesendorf 2022).

These are only a few of the numerous gaps in current energy and climate policy which must be overcome in order for the world to move decisively from the Planned Energy Scenario to the 1.5°C Pathway Scenario. In language which echoes one of the key demands of sections of the global climate movement – 'System change not climate change!' (Beer 2022) – IRENA states that a 'profound and systemic transformation of the global energy system' must be achieved, and with it 'a wholescale transformation of the way societies consume and produce energy' (IRENA 2023: 28). As the report notes, geopolitical developments, principally the war in Ukraine, have thrown up new and unforeseen barriers to such a transformation and caused governments to take retrograde steps, such as new investments in fossil fuel infrastructure (e.g. liquefied natural gas [LNG] terminals).

Despite these caveats, and the many gaps and obstacles which the *World Energy Transitions Outlook* identifies, IRENA argues that it is still possible to achieve a global transition to the 1.5°C pathway. Key energy transition pillars such as physical infrastructure and the skills base of populations must be strengthened and expanded, and policy and regulatory systems which IRENA says are 'still geared toward fossil fuels' must be redesigned to promote renewable energy and reduce emissions (IRENA 2023: 44). Perhaps most significantly, however, IRENA's recommendations contain an implicit critique of what we describe in Chapter 2 as the neoliberalisation of the energy transition and envisage a much greater role for public provision and public intervention in securing the shift to a 1.5°C pathway. IRENA notes that from 2013 to 2020, some 75% of global investment in renewables came from the private sector; however, much of this investment has flowed to 'the technologies and countries with the least associated risks' (IRENA 2023: 26). Thus, according to IRENA, 'stronger public sector intervention is required' to bring about 'greater geographical and technological diversity of investment': instead of focusing on mobilising private capital, climate and energy policy should encourage 'targeted and scaled-up public contributions' (IRENA 2023: 26).

A New State-Centrality?

In a significant departure from the market-friendly language of its *Global Renewables Outlook 2050*, published just two years previously, in 2023 IRENA was advocating for a fundamental shift in the role of the state in order to achieve the goals of the Paris Agreement. The 2023 Inflation Reduction Act (IRA), legislated in the USA after many months of resistance from Senator Joe Manchin, a West Virginia Democrat with strong links to the fossil fuel industry, provides striking evidence of the shift. While the IRA has many critics within the climate movement in the US, who argue that it does not go nearly far enough and makes too many concessions to the fossil fuel industry, both critics and supporters agree that the IRA is 'the biggest piece of climate legislation' ever passed in the United States, by a wide margin' (Climate and Community Project 2022: 1). Although Donald Trump had threatened to repeal it if elected, the IRA represents an explicit shift away from the approach which the Obama administration attempted to legislate in 2009–2010, which would have relied on the pricing of greenhouse gas emissions and the creation of market mechanisms such as a cap-and-trade scheme to reduce emissions over the long term. By contrast, the IRA mobilises direct funding for renewable energy development and foregrounds a much greater role for direct government intervention in order to achieve the goals of US climate policy.

The IRA allocated US$369 billion to Energy Security and Climate Change programmes over 10 years (Democrats 2023). It created a framework for the US to reduce domestic greenhouse gas emissions to around 40% of 2020 levels by 2030 and to massively expand both generation and manufacturing in the renewable energy sector, giving the US a 'competitive advantage in low-cost clean electricity and hydrogen production, infrastructure, geologic storage, and human capital' (Meyer 2022). According to analysis from Credit Suisse, by 2029, 'U.S. solar and wind could be the cheapest in the world at less than $5 per megawatt-hour' (Credit Suisse 2022).

Credit Suisse argues that the official figure of $369 billion which the IRA allocates to spending on climate and energy measures significantly underestimates the actual spending which is likely to occur, because the tax credits through which that spending flows are uncapped. Actual spending is likely to be more than $800 billion, more than double what the Congressional Budget Office estimates as the cost of the measures contained in the Act (Meyer 2022). The Act creates a 'green bank' with starting capital of $27 billion and authorizes the Department of Energy to lend up to $250 billion to finance renewable energy projects (Harris 2022). On the negative side of the ledger, the IRA makes major concessions to the fossil fuel industry, opening up vast swathes of public land to oil and gas exploration and exploitation, and expanding subsidies for CCS (Climate and Community Project 2022).

However, the Act also contains provisions which, according to analysis by researchers at the Climate and Community Project, 'could be a huge aid in stopping the financialisation of the clean energy transition' (ibid.). The current system of tax credits for investment in renewable energy, they argue, has become 'a major tax shelter for private banks and Wall Street', because project developers must go to massive banks like JP Morgan or Bank of America and try to 'sell' their tax break in return for funds. The IRA's direct pay option, they maintain, 'could unleash huge capacity in renewable energy deployment for governments, energy cooperatives, community groups, local business, and nonprofits'.

This is a conclusion broadly echoed by Bryant and Webber in their recently published *Climate Finance: Taking a Position on Climate Futures*. They argue that the IRA is designed to operate 'in a partially green Keynesian manner', steering public and private investment in such a way as to create domestic supply chains for renewable energy technology. In so doing, they suggest, the legislation seeks to bolster public support for spending on climate policy and to demonstrate 'possibilities for doing green industrial policy through and beyond the tools of the de-risking state' (Bryant and Webber 2023: 125–126). This turn to 'green Keynesianism', and the much greater role for the public sector and public intervention advocated by IRENA, implicitly addresses what might be termed the *legitimacy gap* in climate and energy policy. This 'legitimacy gap' is one of the central concerns of this book. To paraphrase the opening words of the popular 1970s science fiction series *The Six Million Dollar Man*, we have the renewable energy technology necessary to bring about a rapid decarbonisation of the global economy, but the 'blockages to doing so are fundamentally cultural and political' (Strauss et al. 2013: 10).

Major energy transitions depend, as a recent study of public participation in energy transitions concludes, on the support of major stakeholders and affected publics (Renn et al. 2020: 3). Our aim in this book is to make those stakeholders and affected publics visible, to understand their experience of transition as it unfolds around them, in the landscapes and regions in which they live, and how and why that experience might lead them to embrace or oppose it. The global trajectory of renewable energy development over the last three decades has created a perception that the energy transition is driven from above, by national states, supranational institutions, and transnational capital. The people and communities most affected by renewable energy development experience it as something happening *to* them, rather than *with* them. This perception can undermine the social legitimacy of the energy transition, and with it, broader social support for climate action.

Approach – The Book Ahead

Our approach builds on the analysis and methodology we developed in our previous book, *Beyond the Coal Rush. A Turning Point for Global Energy and Climate Policy?* In that book, we explored how the legitimacy of the coal-industrial complex was being challenged, both on the ground and at the level of national and transnational climate policy. We followed the contestation of coal mining in three ethnographic case studies in India, Germany, and Australia where local communities were opposing the opening up of new coal mines, or the expansion of existing mines. Based on these case studies, we argued that there was a process of articulation between these local struggles and the larger context of national and international climate policy and movements for climate action. The future of coal, we argued, and with it the future of the planet, was poised at a decisive and historic moment; the power of the coal industrial complex was beginning to unravel with the rapid shift into a new and rapidly intensifying state-renewables nexus (Goodman et al. 2020: xi, 232–234).

This book takes up where the previous book left off. As in the previous book, questions of legitimacy and agency are central. In the previous book, we documented how the legitimacy of the coal-industrial complex was being challenged, bridging the 'gap between climate policy and social action at local, national, and transnational levels' (ibd.: xii). This book employs a similar method, basing our findings on three ethnographic case studies of renewable energy development in India, Australia, and Germany. The shape of the emerging renewable energy system is the object of intense contestation in each of these countries. We approach it from the ground up, through a series of comparative case studies conducted over five years. The book delves into the intricate interplay of policy dynamics and local realities in the renewable energy transitions of Brandenburg (Germany), Karnataka (India), and South Australia. We employ a unique methodological approach, bringing together policy analysis and ethnographic research. By combining these methods, we aim to unravel the multifaceted layers of the socio-political landscape, shedding light on how renewable energy initiatives are conceived, implemented, and experienced on the ground.

The focus of our enquiry broadens beyond that of the previous book to place the local case studies in a regional context, in the new 'energyscapes' and 'energy regions' which are emerging with the expansion of wind and solar energy (Strauss et al. 2013: 11). We expand on the reasons for this broader regional focus in Chapter 2. It is worth reflecting for a moment on a distinctive feature of these 'energyscapes' which may appear obvious, but which is fundamental to the particular problems and challenges which renewable energy throws up.

People cannot live on, or in, a coal mine. As we noted in the previous book, fossil fuel extraction creates 'sacrifice zones' where other forms of human activity are excluded. Renewable energy, by contrast, holds out the promise of co-existence; it is possible, in theory at least, for cropping and grazing to continue on land where wind turbines are installed, and even in some limited form on solar farms. Humans can literally cohabit with solar energy – witness the widespread deployment of rooftop solar in Australia – and they may live close to wind turbines without experiencing the well-documented health effects associated with living close to a coal mine. There is scope, at least, for what two of the current authors have termed 'social co-existence' (Müller and Morton 2021: 65). But as our case studies show, this co-existence is not without tension, opposition, and resistance. While the landscapes in which wind turbines and solar arrays are deployed may be perceived (especially to outsiders) to be spatially empty, uninhabited, or underutilised spaces, they are in fact spaces in which people live, earn their livelihoods, enjoy recreation, and attach value to. The social legitimacy of renewable energy is produced *in* these spaces, and *by* the people who live in them, in a process of dynamic interaction with the policies and priorities of governments and investment capital.

The role of the neoliberal state is crucial to this process. Nation states, we argue, have largely acted as handmaidens to the neoliberalisation of renewable energy, a process we explore in greater detail in Chapters 1 and 2. Thus far, the principal development model of the energy transition, one hitherto legitimised and facilitated by national governments and international institutions (such as IRENA), has been dominated by globalised energy companies building large-scale wind and solar plants, and investment funds seeking 'sustainable' investment opportunities and capturing the income flow from renewable energy. There is investor euphoria for renewables, for upstream 'critical' minerals, and for downstream 'green' hydrogen. As recent legal actions and investigations by media and NGOs have shown, at least some of this euphoria, and the corporate rhetoric that accompanies it, is little more than greenwashing (Carbon Market Watch 2023). On the other hand, however, the International Energy Agency's latest World Energy Outlook predicts that renewable energy capacity is on track to increase two-and-a-half times by the end of the decade – not too far short of the goal of tripling global capacity by 2030 that governments set at the COP28 climate change conference (IEA 2024).

According to the IEA, renewables will overtake coal in early 2025 to become the largest energy source for electricity generation globally; a sure sign that the global energy transition is becoming reality rather than rhetoric (IEA 2024: 15). National and regional governments are competing to attract footloose 'green' finance – seeking a stake in the new sunrise industries. Renewables companies rely on state regulation to facilitate new connections to the grid, to mandate the purchase of electricity from renewable sources, and to guarantee the income streams that underpin profits.

Across Chapters 3, 4, and 5 we attempt to develop a framework that can help us explain what is causing conflicts over the legitimacy of renewable energy; one that draws connections between the processes of contestation that unfold on the ground and what we term the 'neoliberalisation' of renewable energy, in which national states and international institutions work hand in glove with finance capital and globalised energy companies. In our analysis of on-the-ground contestation in the three case studies, we draw on a useful typology of energy conflicts developed by Becker and Naumann (2018) from their own studies of renewable energy development in Germany. Becker and Naumann's typology can help to identify the different kinds of conflicts that may arise around new renewable energy development, what is at stake, and how different actors are drawn into local processes of contestation and deliberation. It has less explanatory power, however, to reveal *why* these conflicts arise; in other words, what is driving them in the first instance. We embed Becker and Naumann's typology in an analysis drawn from political economy. As we argue in more detail in Chapter 1, the current and dominant model of energy transition rests on the *appropriation* of the boundless resources of sunlight and wind and the *accumulation* of capital through the exploitation of those resources. Both these processes have, by and large, been enabled by state regulation, via national and regional governments and institutions.

At a local level, in the places and spaces where wind turbines and solar arrays are deployed, the *appropriation* of wind and solar energy shapes contestations over land, location, and technological impacts; in other words, it defines what is being fought over. The process of *accumulation* creates conflicts over who gets what, who benefits, and who bears the burdens. In other words, it generates new socio-spatial inequalities, or exacerbates existing ones. National, regional, and local governments mediate these processes of appropriation and accumulation through policy, regulation, and procedure. In so doing, they create an arena for contestations over decision-making processes, over whose interests are considered and whose are ignored, who is consulted and who is excluded, and how the identity of a region is defined.

This latter dimension – that of regulation and rivalry – relates to a further fundamental concern of this book, namely the question of democracy. In the journal article which preceded Timothy Mitchell's influential book *Carbon Democracy*, Mitchell observes that 'the democratic machineries that emerged to govern the age of carbon energy seem ... unable to address the processes that may end it' (Mitchell 2009: 431). The 'building of solutions to future energy needs', argues Mitchell, must also involve 'the building of new forms of collective life' (Mitchell, ibid.). IRENA's *World Energy Transitions Outlook 2023* appears to echo this language, stating that accelerating the energy transition 'requires a shift away from structures and systems built for the fossil fuel era'. The transition, write the report's authors, can be a tool with which to proactively shape a 'more equal and inclusive world', but they give little or no guidance as to how this might be achieved.

Yet this aspiration for 'systemic transformation' is expressed not only in policy documents such as the IRENA report; it is also a central concern for the citizens and communities who are living the energy transition, and for whom it is embedded in the fabric of their everyday lives. As Seeta, an activist for climate democracy in the Indian state of Karnataka, put it to us in an interview:

> These stories [of renewable energy] are not any different from fossil-fuel stories, when we go to coal-bearing states or energy hubs of India. Land-issues, water-issues, livelihood-issues, job-issues etc are present there too. How are we going to come up with the vision statement, looking it ... not as a transition but make it into a transformation? (Interview Seeta November 2018).

Our aim in this book is to suggest some answers to the question Seeta raises here: how the energy transition might become a transformation. Our approach seeks to understand the energy transition, which is occurring in different forms, at different speeds, and in different countries as a social and political process. We set out to analyse the changed social and ecological relations which arise in the process of transitioning away from carbon energy. Renewable energy, we argue, is entangled with a wide range of social aspirations and conflicts. It is variously presented as a simple technical fix, a retro-fit of the existing system, an 'energy transition', or a wide-scale social and 'energy transformation'. These agendas reflect the need to conceptualise the *social organisation of energy*, and wider *energy legitimacy*, as critical components in the process of societal decarbonisation under climate change.

Renewable energy is also vested with a variety of utopian hopes: for a distributed prosumer society, for transitional justice, energy democracy, international cooperation, and climate justice (Aranof et al. 2019; Szulecki & Overland 2020). But does renewable energy have the capacity, by itself, to alter socio-ecological relations to create these possibilities?

As Mitchell argues in the Afterword to the 2013 edition of *Carbon Democracy*, 'one cannot predict democratic possibilities directly from the design of socio-technical systems' (Mitchell 2013: 266). In many countries, right-wing populist parties have exploited dissatisfaction with the nature of renewable energy development and sought to build new constituencies around opposition to climate and energy policy (Lockwood and Lockwood 2022). In battles over the shape of future energy systems, Mitchell argues, the question of democracy is open, and open to contestation (ibid.).

The process of contestation, we argue, is highly generative. The contestation of renewable energy development brings about a 're-politicisation' of the energy sector at a local level, where previously the supply of electricity from coal and other sources had been taken for granted (Becker and Naumann 2018: 518). Local actors interrupt and challenge the technocratic imperatives of development. Along the way, according to Becker and Naumann, 'energy provision moves from being the province of technical, legal and business management experts to becoming a field of contestation, in which not only technologies and locations (sites) are at issue, but also a fundamental transformation of the energy system' (ibid.). As the new energy system emerges, so too do new socio-ecological relations. These are often, but not solely, bound up with questions of justice: who wins and who loses, who carries the burdens, and who reaps the benefits. In seeking to exercise agency, to actively shape the ways in which renewable energy development impinges on their livelihoods and life-worlds, individuals and communities undergo a process of learning and make claims which, in turn, force governments and developers to respond and modify the design of the emerging energy system.

The advent of renewable energy, in short, 'socialises' climate change and climatises energy, posing questions of responsibility, agency, and capacity. It plays a key role in advancing emerging social imaginaries in the search for climate agency, and the required ethics of climate-changed life. There is a clear, and what may at times appear an unresolvable, tension between the urgency of decarbonisation and the timescale of systemic transformation. The rapid deployment of new renewable energy necessary to achieve climate goals puts enormous pressure on the democratic processes necessary to secure social legitimacy; governments and renewable energy developers want to speed up approval processes and streamline public consultation, but local communities often dig in their heels.

The process of contestation also challenges, more or less explicitly, what has, until recently, been a central assumption of much climate and energy policy; that renewable capital will clean up the mess that fossil capital has made. As indicated earlier, even transnational policy bodies such as IRENA are beginning to question this assumption. The social legitimacy of 'green capitalism' itself is under challenge on a number of fronts, from movements for climate justice, academic researchers such as ourselves, think tanks, and environmental NGOs. This book can and should be read as a contribution to the critique of green capitalism or 'eco-modernism' and an impulse towards the search for 'new forms of collective life' which might accompany the 'building of solutions to future energy needs' (Mitchell 2013).

The Book

In Chapters 1 and 2 we introduce the conceptual architecture for the book. Three theoretical aspects are identified. In Chapter 1, we outline the socio-ecological appropriation of 'nature's free gifts' of wind and sun for renewable energy, understood as a process of capture, which opens a new 'frontier' in capital-nature relations. We elaborate on the term 'nature's free gifts' originally derived from Marx, and its use in Marxist approaches to ecology and social theory as documented by Saito (2022). Second, we highlight the process of securing a spatial, temporal, and social 'fix' for large-scale renewables, to enable accumulation, and as an emerging aspect of rivalry between region-level authorities and developers to reap the rewards of the renewable energy transition. Third, we focus on the social regulation of renewable accumulation, encompassing state authorities, corporates, workers, landowners, and communities, engaged in a contest to define models for renewable transition and lay claim to 'nature's free gifts'. These three strands are used to develop a conceptual model to interpret the social legitimacy of renewable transition and to guide the comparative analysis.

Chapter 2 tracks the field of renewable energy transition in the three sub-national states where the ethnographic studies are located: Karnataka, Brandenburg, and South Australia. It applies the conceptual model outlined in Chapter 1 to address the full scope of the transition in these regions. The model is used to analytically 'unbundle' the dimensions of transition in the three contrasting regions of study, allowing deeper understanding of the relations in play. The chapter focuses on each state in turn, providing historical and contemporary data about renewable energy policy-making and development at the region level in the context of national authorities and global institutions and agencies. We demonstrate the process of renewable 'capture' by capital, but also how it is contested and the extent to which it prefigures more democratic social relations and new 'forms of collective life'.

In the second part of the book, we discuss the variable experiences of large-scale renewables in localities undergoing intensive development. While policy discourse tends to frame such development as a 'win-win' for investors as well as for climate, there is a more mixed response in host communities where the granular changes to everyday life, environment, and livelihoods can loom large. The three chapters investigate how locals obtain benefits and shoulder burdens at this nexus, and how their engagement with these developments produces new socio-ecological relations. The chapters analyse how local economies and social life are reshaped by the new industry, as new opportunities and tensions emerge among developers, state agencies, and local participants, in the contest to capture wealth on the renewable energy frontier. The accounts look for generative dynamics, that open up new fields of public contestation and pose the possibility of new, more democratised socio-ecological relations.

In Chapter 3, we explore the social relations of renewable energy and everyday life in the Indian state of Karnataka, focusing on the 2 GW Pavagada solar energy park, said to be the largest in Asia, and on the experience of wind energy at the local level. It analyses these installations in the historical context of national and state-level energy policy, framed by wider developmental dynamics and stratification in the Karnataka locality. We contrast the renewable 'resource' with fossil fuel sources and highlight differences between solar and wind power. We discuss the drive to attract renewable investment to the region, along with development finance, in the context of Karnataka's development trajectory. We interpret the transition to renewable energy in terms of social structures, and the extent to which it exacerbates or alleviates pre-existing social divides. In this respect, we see renewable energy, and conflicts centred on it as part of existing social processes, rather than an abstraction from them. There is a strong focus on implications for land, water, livelihood, caste, gender, and environment, including for instance the role, or displacement, of rural landless and lower-caste groups.

Chapter 4 centres on the expansion of wind power and the subsequent 'solar rush' in the German 'energy state' of Brandenburg, where the energy transition (or Energiewende) has been underway for more than two decades. We follow the unfolding process of renewable energy development and socio-ecological capture, paying particular attention to the changing scale of operations exemplified by a move to larger wind turbines and the current shift to large-scale solar farms. We show how the German wind power planning system, designed to avoid or prevent local conflicts, has, at least in some localities, had the perverse effect of intensifying them. We explore the role of local-level planning processes and the imperatives of nature and landscape conservation in the evolution of a socio-temporal 'fix' for renewable investment, and how these interact with corporate investor interests. The chapter provides a rich account of the nexus between a well-established renewables sector and other forms of land use, such as leisure, aesthetics, agriculture, or forestry. The conflict between narratives of regional and local development, prompted, defined, and mobilised in the energy transition, is seen as opening new fields of engagement and disputation in the emerging 'green' economy. As we show, distributional conflicts over the financial benefits of renewable development have led to experimentation with new models of benefit sharing and greater consideration of social ownership.

Chapter 5 focuses on South Australia's Upper Spencer Gulf region in South Australia, which now aspires to 500% renewable energy by 2050. The state has access to world-best onshore wind and solar, with downstream industrial linkages that are now fuelling new spatio-temporal planning horizons. While the state promotes the new energy industry as a 'green' industrial economy, ethnographic research reveals mixed outcomes. Local socio-ecological relations are changing favourably for some groups, such as host landowners and Aboriginal native title holders. Others find themselves left out or further marginalised. Post-construction, renewable energy installations offer few jobs, in localities where unemployment rates are high. Dissatisfaction erupts during the project application processes, where the limits of local demands for meaningful involvement, equitable sharing of benefits, and accountable planning regulation become clear. These, we argue, pose significant threats to the social legitimacy of renewable energy.

The three ethnographic chapters point to a range of comparative and normative insights. These are drawn together in the third part of the book to map the emerging terrain for the social legitimacy of global renewable energy. Chapter 6 draws together evidence-based insights into how renewable energy has been developed in the three regions. The three-part framework outlined in the opening chapter is used to analyse problems of legitimacy in renewable energy development in the three contexts. The three dimensions of appropriation, accumulation, and regulation shape the comparative analysis and underpin a suggested schema for interpreting legitimacy issues in renewable energy transitions. We discuss how progress has been attained, both locally and in terms of the intersecting dynamics of global policy, finance, and advocacy in constituting region-level transitions.

From this, the Conclusions widen the lens to develop a series of substantive recommendations for policymakers, regional, national, and global, who are seeking to strengthen public legitimacy for electricity decarbonisation. It also seeks to draw out implications, in the long haul, for recasting socio-ecological relations under climate change in more democratic directions to realise its fullest potential for societal transformation and democratic engagement. As with energy transitions in the past, the current juncture offers manifold (still undreamt-of) possibilities: we argue for a transition regime that allows for such possibilities to be fostered and realised. There is capacity and agency for distributed renewables, for energy transformations and new forms of energy social ownership and democratisation, in other words, for a 're-commoning' of socio-ecological relations.

References

Aranof, K., Battistoni, A., Cohen, D. & Riofrancos, T. (2019). *A Planet to Win: Why We Need a Green New Deal*, London: Verso.

Asimov, A. (1956). The Last Question. *Science Fiction Quarterly*, 4(5), 6–15.

Becker, S. & Naumann, M. (2018). Energiekonflikte erkennen und nutzen (Identifying and Using Energy Conflicts). In O. Kühne & F. Weber, eds., *Bausteine der Energiewende (Building-Blocks of the Energy Transition)*, Wiesbaden: Springer VS, pp. 509–522.

Beer, C. (2022). 'Systems change not climate change': support for a radical shift away from capitalism at mainstream U.S. climate change protest events. *The Sociological Quarterly*, 63(1), 175–198.

Bryant, G. and Webber, S. (2023). *Climate Finance: Taking a Position on Climate Futures*, Newcastle-upon-Tyne: Agenda.

Carbon Market Watch (2023). *Net Zero of Not Zero? Combatting Corporate Greenwashing, Corporate Climate Responsibility Monitor*, Brussels: CMW.

Climate and Community Project (2022). *Inflation Reduction Act. The Good, the Bad and the Ugly*, Washington: Climate and Community Project.

Credit Suisse (2022). *US Inflation Reduction Act: A Catalyst for Climate Action*, Geneva: Credit Swisse.

Democrats (2023). *Summary: Inflation Reduction Act of 2022*, Washington, DC: Senate Democrats.

Diesendorf, M. (2022). Scenarios for mitigating CO_2 emissions from energy supply in the absence of CO_2 removal. *Climate Policy*, 22(7), 882–896.

Dunne, D. (2023). *Analysis: Africa's Extreme Weather Has Killed at Least 15,000 People in 2023*, London: Carbon Brief.

Goodman, J. et al. (2020) *Beyond the Coal Rush. A Turning Point for Global Energy and Climate Policy?*, Cambridge: Cambridge University Press.

Guterres, A. (2023). 'Secretary-General's opening remarks at press conference on climate', United Nations Secretary-General, 27 July 2023.

Harris, L. (2022). 'Inflation Reduction Bill Uses Public Finance to Stoke Energy Investment', The American Prospect, 2 August 2022.

IEA (2023). *World Energy Outlook 2023*, Paris: International Energy Agency.

IEA (2024). *Renewables 2023. Analysis and Forecast to 2028*, Paris: International Energy Agency.

IPCC (2021). Summary for policymakers. In R. Allan et al., eds., *Climate Change 2021: The Physical Science Basis*, Contribution of Working Group I to the Sixth Assessment Report of the Intergovernmental Panel on Climate Change, Cambridge: IPCC & Cambridge University Press.

IRENA (2021). *World Energy Transitions Outlook: 1.5°C Pathway*, Abu Dhabi: International Renewable Energy Agency.

IRENA (2023). *World Energy Transitions Outlook 2023: 1.5°C Pathway*, Abu Dhabi: International Renewable Energy Agency.

Lamboll, R., Nicholls, Z., Smith, C., Kikstra, J., Beyers, E. & Rogelj, J. (2023). Assessing the size and uncertainty of remaining carbon budgets. *Nature Climate Change*, 13(12), 1360–1367.

Lockwood, B. & Lockwood, M. (2022). How do right-wing populist parties influence climate and renewable energy policies? Evidence from OECD countries *Global Environmental Politics*, 22(3), 12–37.

Meyer (2022). The Climate Economy Is about to Explode. Atlantic Monthly, 5 October 2022.

Mitchell, T. (2009). Carbon democracy. *Economy and Society*, 38(3), 399–432.

Mitchell, T. (2013). *Carbon Democracy. Political Power in the Age of Oil*, London: Verso.

MRIGCCC (2021). *The Remaining Carbon Budget*, Berlin: Mercator Research Institute on Global Commons and Climate Change.

Müller, K. & Morton, T. (2021). The space, the time and the money: wind energy politics in East Germany. *Environmental Innovation and Societal Transitions*, 40, 62–72.

Niranjan, A. (2023). '2023 on track to be the hottest year on record, say scientists'. The Guardian, 8 November 2023.

NOAA (2023). 'July 2023 brought record high temperatures and devastating floods across US', National Atmospheric Administration, 8 August 2023.

Renn, O., Ulmer, F. & Deckert, A. (2020). *The Role of Public Participation in Energy Transitions*, Berlin: Elsevier Academic Press.

Saito, K. (2022). *Marx in the Anthropocene: Towards the Idea of Degrowth Communism*, Cambridge: Cambridge University Press.

Strauss, S., Rupp, S. & Love, T. (2013). *Cultures of Energy. Power, Practices, Technologies*, Walnut Creek, CA: Left Coast Press.

Szulecki, K. & Overland, I. (2020). Energy democracy as a process, an outcome and a goal: a conceptual review. *Energy Research and Social Science*, 69, 101768.

UNDRR (2023). 'New report shows climate disasters are reversing global development at unprecedented rates', United Nations Office for Disaster Risk Reduction, 11 July 2023.

1

The Promise of Renewable Energy

The promise of energy transition is as deep and broad as the sociological and political imagination. Debates about the promise of renewable energy reflect the central importance of energy to society – society is in large part structured by the technology of energy, and how that technology is used and organised. Energy is now intimately intertwined with climate change, through fossil fuel emissions, in ways unimaginable until recent decades. The socio-ecological relations of industrial capitalism have always been climate related: fossil fuels are themselves the product of long-distant climatic conditions. Dependence on burning fuels for energy, from biomass, wood, and charcoal, then to coal, oil, and fossil gas, has always had climatic side effects. These are now in direct collision with capitalist accumulation, signalling the opening of a new geological era, as well as a world-historical moment (Chakrabarty 2022).

The transition to renewable energy both carries the weight of human history and defines the promise for its future. It is hard to imagine anything of greater significance, not just for humankind but also for the diversity of life. Nature, exemplified in climate, is arguably, as Engels famously quipped, exerting its ultimate 'revenge' on capitalism (Engels 1953). Critically, ecology has its own irreducible and amoral autonomy that resists subsumption into capitalism and forces transformations within it. The logic of the atmospheric greenhouse will never bend to the will of capital, even with atmospheric geoengineering. Today climate change poses perhaps the most profound challenge to capital's assumed historical trajectories, forcing a radical challenge to narratives of capitalist developmentalism (Ghosh 2023). The capacity for decarbonisation via renewable energy is at the centre of that challenge.

Global climate policy hinges on the impact and potential of renewable energy: energy from solar panels, wind turbines, and hydropower are viewed as the prime alternatives to the current reliance on fossil fuels. Achieving a reduction in greenhouse emissions should be the first and foremost objective of global climate policy and replacing fossil fuels with renewable energy is the first precondition for attaining it. Other climate measures are ancillary: addressing loss and damage and adapting to the impacts of climate disruption, increasing the carbon 'sink' capacity of the globe, or increasing the efficiency of energy use, all hinge on addressing the root cause of anthropogenic climate change. Only emissions reduction through renewable energy, and reduced energy use and expanded electrification more widely, have any realistic prospect of achieving climate stability. Reflecting this, the vast majority of the 119 countries that submitted their Intended Nationally-Determined Contributions (INDCs) to achieving the Paris goals cited the need to move to renewable energy (UNFCCC 2015b).

Part of the reason for this growing embrace of renewable energy is its growing centrality to economic policy. One of the most remarkable developments in the last decade has been the extent to which renewable energy has become figured as a benefit to society rather than simply as the cost of addressing climate change (including in the IRENA reports discussed in the Introduction). The 2015 Paris Agreement itself points to the importance of the 'co-benefits' of adaptation and 'economic diversification plans' (UNFCCC 2015a: 4.7). The shift is also reflected in successive IPCC reports. In 2014, in its fifth Assessment Report, the IPCC saw renewable energy as advancing development but was concerned about the 'distributive impacts of higher prices for low carbon electricity' (Bruckner et al. 2014: 546). By contrast, the sixth assessment report published in 2022 charted rapidly falling costs for renewable energy, pushing the unit cost of solar and wind power well below the average cost of coal-fired power (Shukla, Skea & Reisinger 2022).

With this there has been a growing emphasis in IPCC reports on new development pathways for emissions reduction and renewables. In 2014 there was a relatively brief discussion of co-benefits in terms of 'Energy Systems' (Bruckner et al. 2014: 544–552). Several aspects were listed – security and resource sufficiency, reduced mining, local employment, water availability, improved public health, and increased energy access. Eight years later, in 2022, the IPCC devoted an entire chapter to the relationship between renewable energy and development, outlining an increased tendency to integrate emissions reduction targets into national-level development planning, and vice versa (in INDCs submitted to the UN). The authors made a direct appeal: 'consider climate whenever you make choices about development, and vice versa'; such planning was seen as essential 'to align towards multiple priorities and achieve greater synergistic benefits'; this pointed to 'a new social contract to address a complex set of inter-linkages across sectors, classes and the whole economy' (Lecocq et al. 2022: 413, 453–454). In sum, the report found 'robust evidence' and 'high agreement' that 'policy integration, addressing multiple objectives, is an essential component of shifting development pathways and accelerating mitigation' (Lecocq et al. 2022: 461).

As renewables move to being the lowest-cost source of energy, the transition from fossil fuel increasingly becomes an economic necessity rather than an environmental burden. Renewable energy development has, as a result, moved to the centre of economic policy, signalling an important shift towards 'green' developmentalism. The new models rehabilitate economic growth as environmentally friendly, to maintain its trajectory and bolster dominant interests. With this come structural limitations. As discussed in this book, renewable energy today is mostly privately owned, structured by the everyday calculus of profit and loss. Private players need a return on renewable capital: this means restricting supply to avoid zero or negative pricing. Maintaining scarcity at times of plenty contradicts the need for abundant lowest-cost renewable energy to reduce fossil fuel use and address climate change. Emissions reduction is thereby made hostage to capitalist social relations.

More broadly for the state, there is the wider need to fuel on-going economic growth, which generally increases energy demand and counteracts aggregate emissions reduction. The process of delinking economic growth from greenhouse gas emissions has long been an objective of climate policy. Policymakers seek evidence of progress in the falling 'emission intensity' of growth, though 'decoupling' at the aggregate level remains elusive (Haberi et al. 2020). Delinking energy from emissions can only occur when emissions decrease more rapidly than overall growth (Foster 2009). To square the circle, as noted in relation to IRENA models, there needs to be substantial slowdown in energy use as well as transition to renewable energy, including the electrification of 'everything'. Yet this comes into direct conflict with economy-wide growth imperatives that rest on expanded energy use.

These dilemmas over how to reduce greenhouse gas emissions and at the same time maintain economic growth have drawn climate policy closer to degrowth and 'sufficiency' debates that focus on the need to reduce overall throughput rather than simply enable delinking. In this respect, reducing emissions entails not just a transition from fossil fuels but also a transition from ever-expanding capitalist growth. Large-scale renewable energy can replicate extractivist ideology and pose its own series of 'resource curses' (Dunlap 2021; Leonard et al. 2022). In particular, large-scale renewables drives deepened inequalities – as Urry argues, 'big energy … is a "great divider"' (Urry 2014: 9). As emission reduction sees ever-larger utility-scale operations these issues are forced onto the agenda, producing a wider shift to new social models and forms of regulation. The shift to renewable energy thereby foreshadows the degrowth agenda (Gills and Morgan 2021): addressing renewable energy is only the start of a longer journey (see Treu et al. 2020; Tokar and Gilbertson 2020).

As we argue in this book, present-day energy transitions can extend the political horizon to address these wider agendas. Our main concern is with utility-scale energy transition. Low-cost renewable energy can produce widescale benefit; the key question, addressed through this book, is who captures that benefit, who bears the costs, and how does this affect public legitimacy for decarbonisation? The answers to these questions shape the trajectory of energy transition, and ultimately, we argue, determine its success or failure. To investigate questions of public legitimacy and renewable energy we focus on subnational regions that have become heavily dependent on wind and solar power, asking what produced the transition, what their transition promises for other regions, what problems transition has generated, and how they can be overcome.

Renewable energy is vested with a variety of utopian hopes, for 'just transitions' (Routledge et al. 2018), 'energy justice' (Finley-Brook and Holloman 2016; Sovacool et al. 2017), and 'energy democracy' (Szulecki and Overland 2020; Wahlund and Palm 2022). It is variously presented as a simple technical adaptation of the existing system, a limited 'energy transition', or a society-wide 'energy transformation'. The different hopes and claims reflect central questions of implementation, of how renewable energy is to be developed. In this, the social organisation of energy and wider energy legitimacy become critical in undermining or enabling decarbonisation (Sareen and Haarstad 2020). Legitimation is intertwined with accumulation (Paterson 2010), as enabling depoliticisation and normalisation, in this case, for corporate-friendly renewables policy. Contestation over changes in land use, struggles over corporate capture, and disputes over multiple other aspects, create legitimacy deficits and force changes in the 'energy regime', pointing to wider transformations (Temper et al. 2020; Newell 2021; Knuth et al. 2022).

As such, energy decarbonisation is a socio-ecological relation, not simply a technical exercise. Again, as Urry argues (2014: 8), sources of social power are shaped by energy structures: 'relations of power depend upon which energy forms are dominant'. As large-scale sources of renewable energy, run by private corporates, have become central to the energy mix, it is increasingly important to enable 'critical analysis and praxis around renewable energy transition' (Knuth et al. 2022: 1007). The agendas are multifaceted and all-encompassing, hinging on questions of inequity and disbenefit, in terms of environmental and labour practices, ownership, and organisational structures, as well as impact on communities and energy users. As noted by the IPCC, addressing these issues forces a broad recasting of the meaning of development itself.

From Socio-ecological 'Fix' to 'Promise'

Privately owned installations offer a predicted profit flow for investors, otherwise they will not go ahead. This requires a set of conditions to be put in place, the 'conditions of production', whether technical, political, social, or spatial, that enable this. James O'Connor argued that limited access to these conditions generates the 'second contradiction' of capitalism, as a constraint on the capacity to accumulate (O'Connor 1998). Such limits are managed by enclosing ecologies for exclusive access to exploit, along with frameworks that enable the capacity to realise returns, and profit. These arrangements reconstitute socio-ecological relations in the name of accumulation and offer a 'fix' for capital, maintaining the required 'conditions'. They establish new forms of accumulation and new frontiers for ecological appropriation, from wind and solar (McCarthy 2015).

In this context the term 'fix' refers to a contingent strategy to ensure a rate of return to capital invested and to do so against prevailing crisis. The concept has been a focus for debate in political ecology and critical energy geography, for instance, with special issues devoted to the topic, such as 'New Energy Scapes' (Bridge and Gailing 2020) and 'Critical Renewabilities' (Knuth et al. 2022). Knuth et al. point out that this often entails 'studying up' to renewables, analysing the emerging configurations and strategies of corporate renewables, as offering an emerging 'fix' for their accumulation problems. Knuth et al. seek more grounded accounts, and in this book we have a similar ambition, with a strong focus on forms of contestation, which we suggest are critical in its development.

The character of the renewable energy 'fix' will depend on relative social relations in terms of capacity to secure cheap labour, access to land and wider ecology, and the state of competition, government regulation, social organisation, and technological development. Any socio-ecological fix for capital though, can only ever be temporary as the process creates new cycles of exhaustion. More immediately, energy transition may deliver for investment capital, but, as Keynes famously pointed out, capital is agnostic on what it invests in – be it a new world war or welfarism or indeed energy transition. If private renewable energy is principally driven by the need for a safe harbour for footloose capital then it is unlikely to be reliable. Whether renewable capital serves the cause of climate action will depend in large part on the state, and on the kinds of transition risk it can impose, what regulatory frameworks it establishes, and on the degree to which it sets a transformative agenda rather than a facilitative one. As McCarthy states, 'a transition would seem likely to require highly active, interventionist, developmentalist states guided by clear industrial policy – a substantial change from the prevailing norms in most advanced capitalist countries at the moment' (2015: 2498).

Arguably the state is recovering the capacity and autonomy to act in the public interests on climate change: as noted, there is already a substantial shift at the national level to more proactive industry policy for transition, away from the blind faith in emissions trading. That shift is linked to the presumed economic benefits of renewables, and hence to the requirement for government facilitation. Importantly, the 'fix' is highly contingent, with today's solutions quickly superseded by new combinations of social organisation and technology. These future possibilities, or 'promises' as we frame them, point to the generative character of capitalist rivalry, that forces technological change and social innovation. The scope of this is only starting to become clear. As McCarthy argues, renewable technology brings 'a transition to profoundly different geographies of energy production, transmission, and consumption [that] could afford powerful political possibilities'; compared with fossil fuels, renewable energy is more 'spatially extensive, dispersed, and visible, and hence more contested and overtly politicized'; this offers, 'in theory … a route towards their greater socialization' (2015: 2499).

As we document in this book, there are signs of such shifts in regional-level state authority, at the 'coalface' of renewable energy transition. Our focus on regional 'success stories' aims to identify the internal contradictions of the current renewable energy 'fix' and thereby point to this immanent potential or 'promise'. A promise is a suitable metaphor for this idea of immanence: a promise is always made in the present with regard to what will happen into the future. We argue the 'promise' of renewable energy arises from the contradictions of green capitalism, which force new previously unimagined agendas and models into view (Buller 2022). The idea of the promise thereby allows both an understanding of the prevailing order and its internal dynamics for change.

The intent here is quite simple: to develop insight into how renewable energy policy can anticipate and move beyond emerging problems. The remainder of this chapter develops a schematic three-part analytical framework for the book. Section 1 explores and develops the underlying concept of capitalist socio-ecological relations of energy. Section 2 outlines three lines of analysis for comparison. The focus is on types of legitimacy conflicts: first, a range of distributional conflicts over accumulation; second, conflicts over appropriation centring on location, land, and technology; third, a series of procedural and identity conflicts over regulation and rivalry. The purpose of the framework is to help draw out what is distinctive about the socio-ecological relations of transition to renewable energy, as against other transitions, so creating the basis for analysis (Eriksen 2016). The chapter ends with a discussion of the combination of political economy and ethnographic method deployed in the book.

Conceptualising Capitalist Socio-ecological Relations of Energy

The eventual replacement of most fossil fuels with renewables signals a substantial shift in the relationship between capitalist society and ecology. Ecological relations change as the replacement of coal by renewables ameliorates climate disruption; equally important, social relations change as the pre-existing society based on the extraction and burning of fossil fuels gives way to a new type of society based on the harnessing of energy from renewable sources. These twin transformations rebound on each other, redirecting forms of technology and associated modes of living. Historically, as with today, the release of society from dependence on one source of energy has created new forms of dependence, reflecting the biophysical logic and social organisation of the new mode of energy. For fossil fuels, we can point to the wide-ranging modes of social organisation that were enabled by coal-fired electricity and that emerged to constitute capitalist modernity (Malm 2016) Industrialism and urbanisation are central aspects, unthinkable without coal-fired electricity. The age of oil brought its own transformations with the combustion engine and the invention of plastics heralding a newly mobile consumer society (Mitchell 2013). The social ramifications overlay the ecological dynamics of extraction and exhaustion, and of course climate change. The currently emergent 'age of renewables' points to a new era, with its own social and ecological dynamics, currently in-the-making as part of the energy transition. The following discussion draws out some of the key aspects of this process, through the lens of 'socio-ecological' relations.

A Process of 'Socio-ecological' Change

In this book we draw on the concept of 'socio-ecological relations' as conveying this combination-in-conflict between society and ecology. It is, in this sense, an expression of the ways in which ecology is continually remade by society, and society by ecology. The relationship can be seen as dialectical, an interaction between socio-structural and bio-physical processes (Moore 2015). Importantly, socio-ecological relations cannot be reduced to a form of dominance, in which capital subsumes ecology. Ecology always has its own autonomy: the carbon cycle for instance is self-generating, producing oxygen from sunlight. Wider ecology and labour as part of it are not produced by capitalism and cannot be reproduced within the capitalist market. Ecology moves with its own rhythms and dynamics, no matter how disrupted by human forces. As a result, and crucially, the economy cannot be disembedded from these conditions of its existence (Polanyi 1944). Reflecting this, the biophysical logic of the carbon cycle and the impact of rising emissions cannot be displaced or repressed forever.

Present-day socio-ecological relations are embedded in the dominant capitalist social formation, which relies upon the appropriation, and therefore degradation, of ecology. The very capacity to produce surplus value depends on combining ecology, energy, technology, and labour power to produce commodities. In this respect, all forms of commodification involve not just the exploitation of labour but also, in the same instance, the appropriation of ecology (Burkett and Foster 2006). Dominant classes reap the benefits, at the cost of ecology and wider humanity. These classes depend on the process of appropriation that leads to degradation: the fossil fuel 'majors' being a case in point. The climate crisis points to a different form of society, embedded in the requirement to be free from fossil fuels. This may in practice become a post-capitalist society, but it is certainly a different form of capitalism, driven by different imperatives.

Capitalist commodity production has always involved what Marx called 'original accumulation', that is, the violent expropriation of ecological attributes – energy, materials, and wider ecologies, often from other people (Marx 1976: Pt 8; Angus 2023). As local ecologies become exhausted, new frontiers for extraction need to be opened up, with new technologies of production and forms of social organisation. Ultimately, though, continued extraction erodes the ecological foundations for capital accumulation and undermines planetary capacity (Foster 2009). This process or appropriation centres on 'capitalisation', where the state licences the 'resource' to be appropriated, integrating it to the dynamics of 'self-expanding' capital (Rosewarne 2022).

The capitalist pursuit of extraction faces the limitations imposed by both ecology and technology. The rivalry to accumulate drives technological development to overcome ecological barriers and create new socio-ecological relations. These create new forms of appropriation and whole new industries, often with a long-term investment lock-in, and with various intended and unintended impacts. The move from windmills, watermills, and canals, for instance, to steam-driven machinery allowed a leap in production, but at a huge cost from coal mining, coal burning, and later climate change (Freese 2003). The move from household coal to thermal electricity and then to oil dependency had similar side effects.

To emphasise: capitalist development and ecology are driven by fundamentally different, and clashing, imperatives. Capitalist society is driven by the necessity for an ever-increasing expansion to maintain the viability of capital. Ecologies are driven by their own processes of biophysical transformation. Both sets of drivers exist autonomously but are bundled together into different contradictory combinations, with energy production and use depending on the state of technology and productive forces (Moore 2015). With this perspective of unity-in-separation, we can interpret socio-ecological relations as dynamic and interactive. With this, we argue against the common assumption that the capital-ecology relationship is a zero sum, viewing ecology simply as a fixed stock to be used, diminished, or restructured by capital. The socio-ecological relationship is much more dynamic as ecological limits are created by capital and by its destructive extraction, not by ecology itself. As such, the trajectory of accumulation is shaped by ecology and its breakdown: in this sense, ecological crises stem from the 'production of nature' under capitalism (Smith 1984), climate change being the archetype. To reference Engels again, nature escapes human 'mastery' and wreaks its 'revenge' (Engels 1953: Pt IX):

> Let us not, however, flatter ourselves overmuch on account of our human conquest over nature. For each such conquest takes its revenge on us. Each of them, it is true, has in the first place the consequences on which we counted, but in the second and third places it has quite different, unforeseen effects which only too often cancel out the first. Thus at every step we are reminded that we by no means rule over nature like a conqueror over a foreign people, like someone standing outside nature – but that we, with flesh, blood, and brain, belong to nature, and exist in its midst, and that all our mastery of it consists in the fact that we have the advantage over all other beings of being able to know and correctly apply its laws.

Renewable energy releases capital accumulation from the constraints and obstacles of fossil fuel extraction, just as fossil fuels release accumulation from the constraints of horsepower, bioenergy (wood, charcoal), watermills, and windmills (Malm 2016). The scope of its impact is only limited by the scope of social power relations, infrastructure, and technology. The abundant foundations for renewable energy create new agendas for socio-ecological transformation, expressed for instance in 'green new deal' debates and the wider agendas for more active state engagement and for a wider 'energy commons' (Barry 2021). Such agendas, we argue, gain traction as the basis for extending the legitimacy of energy transitions into the future.

However, there are constraints. Renewables must appropriate the 'free gifts' of nature, and not just wind or sunlight, but also space for installation and transmission infrastructure, as well as minerals for its machinery. Such appropriation is central to the logic of accumulation: Saito puts it simply in his *Marx in the Anthropocene*, 'natural forces are thoroughly exploited by capital as "free gifts" … nature is free, and capital seeks to utilize its power as much as possible' (Saito 2022: 223). Along the commodity chain, disputes among capitalists, workers, and the broader community are commonplace, and many of these conflicts arise from the reliance on large-scale corporate-run utilities (Gazmararian and Tingley 2023). This dominant way of organising renewable energy is based on the model of centralised generation and reflects the continuing power of existing energy corporations and existing institutionalised expectations, knowledge, infrastructure, and regulations. Distributed renewable energy challenges the model, as exemplified in the rapid take-up of rooftop PV solar generation, but remains subordinate.

In this book, we focus on the dominant model of utility-scale production and how its problems may be overcome. The state is critical in this: the key driver of decarbonisation is not direct climate change impacts and 'physical risk', but rather the indirect 'transition risks' arising from state-led climate policy and legislated fossil fuel phase-out (Goodman 2022). Climate change may be an imperative at the global level, but mounting costs of climate change are not necessarily being absorbed by capital to the extent that the physical impacts of climate change are enough to force changes in overall accumulation models (Foster 2009), despite acute pressures in sectors such as insurance. Government, which attends to the general populace and not just the capitalist elites, is indispensable in forcing the agenda. Even then, fossil fuel corporates respond by accelerating mining and drilling to extract 'value' before assets are 'stranded' (Sinn 2012), and energy investors can seek hedging 'positions' with a foot in both camps, conjuring fears about energy intermittency.

Yet, large-scale renewable energy is now adopted almost universally by national states as the main way to achieve energy decarbonisation. The state realigns energy systems to supplement fossil fuel-based energy systems and ultimately to replace them. It takes a leading role, instigating the transition in the name of climate action (Kern and Rogge 2016). In doing so it acts in the interests of 'capital in general', as a 'collective capitalist' beyond the narrow interests of the fossil fuel lobby. The scope for this limited autonomy is critical in shaping the possibilities for decarbonisation: in countries and regions that are not dependent on the extraction of coal, gas, or oil, the fossil fuel sector can wield relatively less power and this can help explain the emergence of renewable energy frontrunners: they have relatively weaker fossil fuel vested interests and also have a direct incentive to embrace renewable energy in order to reduce fossil fuel imports. In aggregate though, state capacity, at least at the scale required, is still lacking. Across the globe, governments have been unwilling or lacking capacity to act in society's wider interests to advance the renewables transition at anything like the pace that is required, despite the sometimes heartfelt declarations from governments or the UN on the urgency for action. As tracked by the annual 'Emissions Gap' reports, national state 'contributions' to meeting global emissions reduction goals remain inadequate (UNEP 2023). There is evidence of some progress and growing autonomy from fossil fuels, though this remains limited. It was only in 2023 for instance that the annual Conference of the Parties to the UNFCCC was able to make the call for 'transitioning away from fossil fuels in energy systems' (UNFCCC 2023: 28.d) – itself a conscious step back from calls to 'phase out' fossil fuels in the interests of compromise.

To be capable of achieving the process of 'transitioning away' and advancing renewable energy, the state needs to act 'above' fossil fuel interests. Driven by the need to overcome the climate crisis and to embrace the emerging benefits of renewable energy, it seeks out renewable energy models to overcome the growing crisis caused by fossil fuel energy but at the same time enable continued or expanded capital accumulation. The models are implemented 'from above' as part of a state-led transition that creates and entrenches new energy interests and elites. The process is analogous with the 'passive revolution' identified by Antonio Gramsci, where the crisis forces states to develop a new framework for accumulation that is then enforced against contenders (see Morton 2010; Carroll 2020; Ford and Newell 2021; Gill and Morgan 2021). Gramsci's 'passive revolution' saw the rise of fascism to resolve the economic crisis of laissez-faire capitalism; arguably in the current period we are seeing the imposition of 'green' capitalism to overcome the ecological crisis of neoliberal capitalism (Buller 2022). This shift is ongoing, open-ended, and subject to its own legitimacy challenges: outcomes are not preordained and are highly contingent. Carroll for instance posits new frames for 'energy democracy' as offering scope for 'non-reformist reforms', opening up more transformative possibilities in the sense suggested by Andre Gorz (Carroll 2020; see also Bond 2020; Stuart et al. 2020). Mookerjea goes further, arguing for 'regenerative delinking' beyond the productivism of energy transitions imposed by 'green' capital (2019). Our stress is on renewable energy as generative, producing new possibilities, both inside and beyond energy transitions, as a way to gain critical perspective on a dynamic process.

Undeniably, energy has entered a period of flux and transformation – questions of strategic planning, foresight, and capacity for policy anticipation have become central. In this context, what forms of social organisation become necessary – and possible? What forms of social struggle emerge and how do they create legitimacy problems and change power relations? To what extent can we 'read off' upcoming social and political possibilities from these tendencies? The renewable energy transition has inherited many of the institutional and infrastructural conditions of fossil energy systems. These have been subject to neoliberal market 'reforms' over several decades, that have benefitted large energy oligopolists, which are now positioned to capture renewable energy systems. Yet, as outlined in the introduction, renewable energy can be organised in ways that do not simply replicate the existing structures and create its own pressures for socio-ecological change, even for transformation. Legitimacy conflicts are central to these dynamics.

Renewable Energy Legitimacy Conflicts: A Framework

The socio-ecological relations of renewable energy have interacting economic, political, and ecological aspects and produce a range of legitimacy conflicts. There are conflicts over climate policy and the extent to which transition to renewable energy enables decarbonisation or simply escalates energy use. Energy transition, though, does not simply concern climate. There are multiple other dimensions, reflecting the centrality of energy to society. These translate into multiple allied co-benefits of transition, from low-cost energy to health benefits, energy security, empowerment, and access, all of which are central factors in enabling the legitimacy for national transitions. These aspects, and others, come into play in the debate about energy transition, reflecting the broad socio-ecological power relations it invokes, including and beyond the climate antagonism. As Sareen and Haarstad observe, 'it is of critical importance that reforms and policies aimed to stimulate sustainable energy transitions address power imbalances as well as CO_2 emissions (2020: 49).

We ground our analysis of renewable energy legitimacy conflicts in three key aspects of the transition process. First, we emphasise capitalisation, in terms of where and how ecology is appropriated. Second, we focus on accumulation, where extracted energy is commodified to generate returns. Third, we discuss state regulation and rivalry and how it can secure the overall 'fix' for renewable capital. These aspects are seen as generating legitimacy problems, antagonism, and agendas for contestation. The correlation is interpreted as a process of 'articulation', using the concept developed first by Althusser to correlate underlying social crises and ideological formations and later by Stuart Hall to interpret relations between capitalism and racism (Featherstone 2011). Hall emphasises the contingent character of articulation and the extent to which it is grounded in particular places and enacted as part of a defined temporal narrative. We have found this concept of spatio-temporal articulation particularly useful to help conceptualise the relationship between the underlying climate and energy crisis and the emerging legitimation issues for renewable energy. We focus on the spatial context of the region and map renewable narratives of transition over time, highlighting the legitimacy issues that emerge. The approach allows for a focus on relationality – and gives critical purchase on what may seem a miasma of multiple and contingent aspects, simply assembled or bundled together.

In analysing the range of legitimacy conflicts over renewables, we draw heavily on the work of Becker and Naumann, who develop a typology of contestations in a number of different renewable energy settings in Germany. They argue we need to clarify who is involved, what is being fought over, and in which arenas the conflicts play out (Becker and Naumann 2018: 512). Several conflicts are identified. Locational conflicts emerge when the use of land for energy initiatives is contested, along with concerns about the resulting impact in terms of livelihood or for the landscape. 'Fundamental' technological conflicts are rooted in disagreements over the acceptance or rejection of specific energy sources, such as coal or nuclear power, and also over the form and scale of these technologies. Perhaps most important are the distributional conflicts, where debates centre on who reaps the financial rewards of new renewable energy projects. Procedural conflicts involve disputes over planning processes, decision-making procedures, and issues of access to information. Last, identity conflicts encompass perceptions of a region and its distinctive identity, whether framed as an energy hub, tourist destination, or health-focused locale (Becker and Naumann 2018: 511–513).

We take this typology and ground it in three underlying contradictions of energy supply. First, we address the problems of legitimising the appropriation of land and environment for large-scale, 'high-tech' renewable energy installations. These problems reflect an underlying contradiction between accumulation and ecology, framed as an aspect of wider nature-capital contradictions. Second, we focus on the distributional problems created by accumulation itself, where renewable energy corporates are rewarded at the expense of other players. This reflects conflict between capital and labour, where surplus accumulation hinges on exploitation. Third, we find a set of regime-level problems, with contestation over procedural legitimacy and the process of regulating renewables and defining development pathways. Here there are powerful drivers in terms of rivalry within and between regions, in the context of footloose renewables finance. As outlined in what follows, these three interrelated aspects can be used to investigate the legitimacy conflicts identified by Becker and Naumann.

Locational, Land, and Technical Conflicts Over Appropriation

The corporate 'rush to renewables' poses new questions about processes of appropriating wind, sun, and waterflow. Wind, solar, and hydro are renewable sources of energy, but the technology that captures this energy is not renewable (or not yet). Renewable plant and technology is produced from metals, such as iron and steel, copper, nickel, manganese, chromium, zinc, silicon, and rare earths, that are exhaustible. They produce forms of waste that (again, as yet) are difficult to reuse, recover, or recycle. These supply chain issues for renewable energy can be reorientated to deliver new industries, in recycling, for instance. Metals production itself can be fuelled by renewables, creating options for a renewable circular economy. Transition plans for coal regions often involve new roles in producing steel and other industrial requirements for the renewables sector. For example, in Australia, proposals for 'green steel' envisage a positive feedback loop between renewable energy and the steel sector that could employ large numbers of workers in formerly fossil fuel-dependent regions. Here, for instance, it is hoped renewable energy, as hydrogen, will replace metallurgical coal in the steel-making process, driving demand for renewable energy and providing employment for ex-fossil fuel workers (Wood and Dundas 2020).

Furthermore, installations and infrastructure are themselves place-fixed assets, that require access to land, roof tops, or seabed, and can have important impacts on local ecologies and livelihood. The reorganisation and appropriation of ecological space at the very least drives a restructuring of existing patterns of land use and access. It can also deepen or produce new social inequalities, sparking contention over planning approvals and zoning procedures, questions of scale and extent of the development in the particular locale, changes in ownership, possible displacement of existing uses and employment, for instance in agriculture, along with spatial allocation and extension of the electricity grid. The process of establishing the new ways of appropriating ecological forces, and opening up new frontiers in accumulation, has the effect of (re-)industrialising regions, with related effects (see Backhouse and Lehmann 2020). While renewables may be embraced as a means of restoring ecological and climate stability, they also disrupt some or other peoples' relationship to ecology, local landscapes, and sources of livelihood.

The spatial logic of renewables appropriation is important: it is driven by the biophysical requirements of renewable energy, and this can involve a spatial shift in energy production away from the existing industrial heartlands that historically have borne the brunt of coal dependence (see Cass et al. 2022 for an Australian example). Coal regions may suffer economically, losing employment and livelihood without a commensurate boost from renewable energy. The transition for such regions is therefore not necessarily from coal to renewables but from coal to a diversity of other economic activities. By the same token, new renewable energy zones tend to emerge on 'greenfield' sites, or previously agricultural sites, often on a large scale, far from centres of population and from industry. Questions of 'just transition' at the local level can be as intensely posed for these regions hosting renewables as for regions losing their existing fossil fuel industry. It is clear there is no 'friction-free' process of appropriation. At the same time, it is also clear that some forms of appropriation are more damaging, and at different scales and contexts, than others, and are contingent on the way land, resources, and labour are organised in the transition (Pearse and Bryant 2022). The contrast with the extraction and burning of coal for electricity, always highly damaging, is worth keeping in the picture.

The developing technology of renewable energy can pose very different scenarios for the process of appropriation. There is variation in the process of acquiring land and enclosing it, with wind power more able to coexist with other land uses than solar power, for instance. There is variation in scale, from household to large utilities, there is also variation in type, for instance, in terms of agri-solar as against conventional solar and self-cleaning panels as against panels that require water. There is also variation in geographical logic of development, whether on the flatter low-lying plains for solar as against the foothills and ridges for wind turbines. These developing technologies can become an important focus of concern in terms of which form of renewable energy is favoured over others. Energy storage is another case in point, in terms of the difference between battery-based storage and pumped hydropower, for instance. The social legitimacy for renewable energy is profoundly shaped by the ways in which these technological and locational issues, as Becker and Naumann characterise them, are defined, contested, and readdressed.

Distributional Conflicts over Accumulation

Many of the key conflicts over renewable energy centre on distributional questions, of who gets what. Arguably all of society can benefit from renewable energy insofar as it enables reduced greenhouse gas emissions. The society-wide benefit is often mobilised to legitimise private sector interests and obscure the skewed distribution of costs and benefits in renewable energy development. As noted, renewable energy is currently constrained by capitalist 'business models' aiming to capture sufficient returns (Christophers 2024). The renewable energy sector is capital-intensive and highly mechanised and has built-in economic incentives to become further so. Corporates seek to minimise marginal costs and co-benefits such as the provision of locally based, well-paid, and rewarding jobs can only undermine returns. Much renewables policy is now geared to attracting and assisting larger companies, and to favouring their interests over small and non-corporate players, further undermining possible region-level benefit.

Across the renewable sector, there is a rapid process of globalisation and concentration, with larger companies moving to establish ever-larger-scale utilities. The concentration of ownership directly affects the distribution of benefits for local communities and can lead to sharp contests between the interests of a locality and those of large, relatively distant corporations that establish and run large renewable energy utilities. Yet, against market capture, there is considerable experimentation and innovation in seeking new uses for rapidly changing technologies. Renewable energy has unleashed a process of market disruption, upending the hierarchic and centralised structure of power provision. In the process, there may be extensive possibilities for cumulative societal 'take-up', allowing the sector to realise some of the wider socially transformative modes of energy 'from below' (Rosewarne 2022). At the same time, this volatility and uncertainty generate strong 'rent-seeking' pressures for the larger players. They respond by seeking to insulate their investment and income flow, for instance via long-term energy supply contracts (Bridge et al. 2020; Baker 2022).

Probably the most important driver of conflict over distribution is the systemic logic of accumulation that requires ever-increasing throughput of extracted material and expansion of energy use in order to expand profit and growth. Distributional concerns in renewable energy principally arise from the social organisation of renewable energy development, and in particular from the extent to which it relies on private operators, which in turn rely on accumulation. With this we can view distributional contests as dynamic and generative, as each agreement in the realm of accumulation leads to new claims, escalating demands. As the renewable energy sector seeks returns within traditional capitalist frameworks, questions about the quality of jobs, their localisation, and the impact on communities become central. The sociological lens further underscores that the process of accumulation has implications not just for economic growth but for the broader social fabric, creating constituencies and contestations in a dynamic process.

Procedural and 'Identity' Conflicts over Regulation

Finally, there is the question of legitimacy conflict over decision-making and over the process of defining regional development identity. As previously noted, the state is the central arbiter of access to ecology's 'free gifts': it is the state that sanctions appropriation and therefore enables the capitalisation of ecology. The right to extract energy from wind, sun, and water flow can be priced through licencing regimes, as with mining, though in practice governments provide incentives for renewable energy while also imposing levies to ensure wider societal benefit.

Energy supply is treated by state authorities as a privileged matter of 'national interest', regardless of whether it is in private or public hands. Whole government departments are devoted to the task of administering to the requirements of the energy sector, with corporatised energy entities, whether state-owned or private, treated as part of the 'commanding heights'. These state authorities are often deeply intertwined with fossil fuels as the principal legacy mode of energy supply. The process of establishing new structures catering to the profoundly different modes of electricity supply and consumption under renewable energy has proved to be one of the major barriers to realising the potential of renewable energy. That challenge is (at least in equal parts) a political and a technical challenge, a matter of legitimacy as much as of technology.

Despite the evident pressures against it, renewable energy has gained a degree of political traction and a key aspect of this is the mobilisation of state power to realise the broader socio-ecological co-benefits of transition, notably in terms of industrial development but also in terms of energy security and population health. With renewable energy now substantially cheaper than fossil fuels and given its cited benefits for the development of allied forms of technology, innovation, and manufacturing, especially with the posited electrification of society, governments have begun to compete to attract renewable investment. Rivalry to attract the substantial capital flows now entering the renewable energy sector has seen governments establish a variety of initiatives, agencies, and departments to encourage the influx.

The regulatory outcome is deeply uneven and contradictory, with some countries acting as 'laggards', others as 'leaders', depending on the variety of pressures coming into play in national climate and energy policy. At least for the 'leaders', including the regions analysed in this book, we can begin to speak of an emerging renewable energy 'industrial complex' with a substantial reorientation in institutional power. As the renewables transition has become an inevitability, the fossil fuel sector has sought to diversify from their incumbent position while maintaining a role for fossil fuel generation as a bridging fuel, or as power in the interests of energy system 'stability'. There is evidence of experimentation in the modes of social organisation of renewable energy, although, as noted, the dominant trend is for state-facilitated utility-scale private provision, generally for pre-existing energy corporates, though this may change with developing technology and the changing social organisation of renewable energy production.

The reorientation in energy provision is closely linked with development strategy, and this rests on public legitimacy. State regulation of renewable energy proceeds in ways that reflect and absorb public concerns about energy access and affordability. This process closely approximates Becker and Naumann's emphasis on procedural and participatory engagement with renewable energy as key components of social legitimacy for energy transitions. These aspects need to be understood as generative – as conveyed in the concept of transition itself, with nothing set in stone. New constituencies and political forces have become evident in the renewables transition, and in tandem with the new more distributed modes of social organisation in renewable economies, we may yet see related transformations in the modes of democratic participation and political institutions.

Certainly the process is highly dynamic. Decarbonisation remains contested at every phase of renewable energy development and is producing new modes of regulation and provision. The evolving role of the state in shaping renewable energy policies is not just a technical or political challenge but a transformative and indeed cultural process with broad social implications. The shift to private ownership of renewables reflects the assumptions of the neoliberal energy transition. This is reflected across the three case study contexts. In India the renewable energy generating sector is almost wholly privately owned, in contrast with the largely state-owned fossil fuel sector. In Australia, electricity generation overall (apart from rooftop solar) has largely been privatised, while other aspects of the power system remain in public hands. Germany has a tradition of municipal and cooperative renewable energy, although this has been threatened by a move to tendering out in a renewables sector, which is increasingly dominated by the large private players (Jungjohann and Morris 2016).

The neoliberal narrative suggests the onwards domination of private interests, but this is belied by the simultaneously growing role for state authority in enabling transition. Crucial in this is the issue of legitimation. State involvement in the renewables transition is always a reflexive process, where the regulation of renewable energy is redirected in ways that reflect or anticipate the need to enable social legitimacy and capture the public imagination. With this we can see the state actively intervening in the construction of regional and local identities, integrating local development aspirations with forms of progressive ecological modernity, delinked from fossil fuels.

The emergence of this cultural and industrial 'complex' around renewable energy enables a shift in political constituencies and a reconfiguration of democratic participation, and in turn opens up new forms of democratic agency and modes of state action. In this there is a close dialectical relationship between renewable energy commodification and the assertion of an energy commons. Polanyi tracked how commodification, under the laissez-faire regimes of the late nineteenth and early twentieth centuries, generated a socio-political response in terms of an impulse for social protection through the state (Polanyi 1944). Socialisation and re-commoning are defined by their antagonism to commodification and therefore to private and corporate interests (see Goldman 1998; Laxer and Soron 2006). These dynamics are expressed in many of the contestations we analyse, and in the claims and agendas that emerge. Under neo-liberal precepts, private capital is being invited to establish the new renewables sector and this is generating deep conflicts and producing new agendas for re-socialising and (re-)commoning energy. A key aspect of this is the underlying public purpose of decarbonising energy – as a defence of the planetary commons, of enabling conditions for both human and wider biophysical planetary survival (Rockström et al. 2024). Where private sector renewable energy fails to address this and instead pursues its own accumulation agenda, we can expect re-commoning.

Significantly, as the state acts for emissions reduction, seeking to protect society from the growing climate crisis, it lays the foundations for energy commoning. The process of emissions reduction is already leading to new roles for state authorities, and especially so in the energy sector (Haney and Pollitt 2013; Steffen et al. 2022). Energy transition is actively enabled and catalysed by the state, to the extent of state agencies arranging land acquisition, establishing transmission infrastructure, and guaranteeing returns. The 'plug and play' installations that are becoming increasingly prevalent can be described as private sector operations only in terms of the generation infrastructure, whether wind turbines or solar panels, and even these may be enabled by large incentives from state agencies. This lays the basis for new claims, especially when the private sector fails to deliver. As Huber points out, rather than vest hopes in the emergence of a more distributed energy system, 'it is more convincing to imagine social transformation through our existing centralised power grid' (2022: 251). He argues the existing grid and utility-scale renewable energy installations are already effectively socialised, and that properly understood 'the existing electric utility industry is a gargantuan form of socialised production' (Huber 2022: 254). As we demonstrate in the following case studies, the desire for energy commoning is a key shared characteristic of contestations over renewable energy, including the demand for a state-led civil commons in energy as a 'right to energy' perhaps, along with extensive social ownership.

Overall, the three aspects described here – appropriation, accumulation, and regulation – embody contradictory forces that drive the antagonisms and define the transition narrative. Ecological appropriation enriches elites but also degrades ecologies; accumulation comes at the expense of subordinated classes; and regulation can enable domination over both people and nature. The shared imperative to challenge the exercise of elite power in these domains is what unifies the contestations. Addressing these interrelated aspects, as we have argued, reveals key policy challenges for navigating the energy transition.

Overall Schema: Legitimacy Conflicts in the Transition

There has been a growing literature on 'social acceptance' for renewable energy (Gross 2007; Ribiero et al. 2011; Fast 2013). A special issue on the topic was published in Energy Policy in 2007 that identified socio-political, community, and market acceptance as key aspects (Wüstenhagen et al. 2007), followed by another in the same journal in 2023 emphasising the influence of time, power, and scale in shaping the acceptance of decision-making processes around renewable energy (Ellis et al. 2023). Socialised participation and ownership in distributed renewable energy is seen as encouraging acceptance, along with regional planning, as part of a wider vision for development and sustainability (Wolsink 2012). In analysing transitions, ideas of 'social legitimacy' are integral, and we argue can be more sociologically incisive than the more familiar formulations of 'social acceptance' (Wüstenhagen et al. 2007), 'energy governance' (Michalena and Maxwell Hills 2013), or the 'social license to operate' used in the mining sector (Thomson and Boutilier 2011). As Paterson argues, following James O'Connor, structures of 'legitimation' are critical to enabling accumulation (Paterson 2010): legitimation obscures contradictions in the accumulation model, often subsuming private interests as ostensibly serving public benefit, for instance in this case, from emissions reduction. By the same token, de-legitimisation offers the possibility of making social power relations visible, politicising fields of energy transition, making them objects of policy, and of contestation and transformation. Clearly, energy transition arises within ongoing political and cultural struggles, involving a variety of social groups and issues, across livelihood and ecologies (Hughes 2004;

Table 1.1 *Renewable energy and legitimacy – summary schema*

Legitimacy conflicts	Underlying process	Aspects	Sites
Locational, land, and technology conflicts	Appropriation (of ecology)	Ecology/land, as a 'resource', mode of extraction	Land acquisition/use, mode of installation
Socio-spatial distributional conflicts	Accumulation (of capital)	Commodification, finance, labour, and livelihood	Supply contracts (Power Purchase Agreements), jobs, and pricing
Procedural and regional identity conflicts	Regulation/ rivalry (by state authorities)	Decision-making and regional development framing	Planning bodies and development Ideology

Smith and High 2017; Radtke and Scherhaufer 2022). As Mitchell argues, renewables are defined by highly contingent 'battles over the shape of future energy systems' (2013: 238, 267). Existing energy regimes are entrenched, with minimal capital costs: new energy technologies require up-front infrastructure and can bring unpredictable effects, not least in terms of social exclusion (Unruh 2006; Araújo 2014; Bainton and Kemp 2021). Today, such battles are embedded in the 'negative value' of the current fossil fuel energy regime, in terms of the greenhouse gases it produces, and resulting climate change (Moore 2015: 277). Entrenched fossil fuel subsidies are a barrier to social equity and sustainable development goals in the energy transition (Rentschler and Bazilian 2017). Escalating emissions reduction targets introduce new and dynamic elements into the process. With this we see a close articulation between the underlying contradictions generating the climate crisis and the emerging energy transitions.

The three analytical aspects described earlier – appropriation, accumulation, and regulation/rivalry – each entail questions of social action and legitimacy. We argue that navigating the dynamics of renewable energy transition requires a holistic exploration that can encompass structural change and social legitimacy, and how they are articulated. With this we are able to recognise transitions as having a generative effect, producing agency, legitimacy, and learning. The three-part analytical frame deepens the analysis of legitimacy conflicts and the arenas in which conflicts unfold (Becker and Naumann 2018: 512). As outlined in Table 1.1, issues of appropriation encompass disputes over energy technology and also locational disputes involving siting, land use changes, and ecological impacts; questions of accumulation encompass issues of distribution, both in financial and social terms; issues of regulation and rivalry entail procedural debates regarding participation opportunities, as well as tensions related to regional or cultural identity. While not exhaustive, we posit these points of contestation as critical junctures in the trajectory of decarbonisation pathways within the renewable energy capitalisation process. The table illustrates the three legitimacy conflicts for analysis, drawing out aspects and key sites as discussed in the case study chapters that follow.

Legitimacy conflicts exert a generative force over time, shaping the socio-ecological fix for renewables. The temporal aspect is critical, yielding narratives of transition showing how new constituencies emerge, altering the political landscape. A reflexive temporal dimension emphasises that each point of contestation is not static but part of a learning process over time, creating narratives of renewable energy development where agreements and conflicts lead to the emergence of new claims and understandings. Using the three-part framework, we seek to make the connection between these questions of social legitimacy and the underlying dynamics of agency and structural transformation in the transition. In this we aim to encompass the reflexive and generative nature of legitimacy struggles, the transformative potential of transitions, and the dynamics of agency and learning embedded in the contested process of decarbonisation.

The challenge here is as much to synthesise the shared challenge, defining the common field of contestation, as to iteratively identify challenges as they emerge. There may be multiple challenges to an energy regime, but arguably they only have purchase when correlated to generate a wider political vision. Here the concept of 'assemblage' can be useful in signifying the emergence of a common field for social agency in the development of renewable energy. 'Assemblage thinking' can be used to conceptualise inter-relations between ecology and society (see Albert 2022) and to rethink the spatial and material contexts for ethnographic engagement (Ghoddousi and Page 2020). Arguably it is most useful for investigating agency, as a process of arrangement across different aspects, as reflected in the original French meaning as *agencement*, 'to arrange, to lay out, to piece together' (as against assemblage, 'to join, to gather') (Nail 2017: 22).

A focus on how agency is pieced together is critical in any political context. Here we suggest there is a clear articulation between underlying contradictions in the development of utility-scale capitalist renewables and the legitimation struggles that arise. This articulation generates a field of contestation, across both material and non-material dimensions, widening the scope for agency beyond human-controlled factors (see Marcus and Saka 2006; Armstrong 2021; Hoffman et al. 2021; Helfer 2022). 'Nature's free gifts' are never free: the energies of wind and sun can only be appropriated in space and time, in the concrete social, historical, and biophysical settings, which in our case studies shape the energy landscapes of Karnataka, Brandenburg, and South Australia. The social settings in which energy conflicts then unfold are also material settings, and vice versa, the material energy landscapes of wind and solar enable new social formations, and are interwined in the 'social making of arguments' (Reusswig et al. 2016: 226).

A focus on the dynamics of assemblage across the resulting contestations can be very revealing in terms of how initiatives are co-inspired and co-related to produce an effective political agenda (Featherstone 2011). Quoting Doreen Massey, Featherstone suggests the approach allows 'a focus on the "enablement of new relations"' in particular spatial contexts, and how they are 'generated through the coming together of previously separate or disparate trajectories' (2011: 141). This draws attention to the process of creating relationalities, solidarities, or alliances, not as automatic or assumed but as 'reworked, constituted and brought into contestation through the ongoing conduct of politics' (2011: 142). Through this process, we argue, commoning increasingly defines the scope of the challenge, across the broad 'assemblages' that encompass it.

Approach

How to materialise the socio-ecological relations of climate change? There is a vast gulf between the biophysical global-scale process of climate change and the mundane questions of everyday decarbonisation and energy transition. This is the central problem for climate action and has driven our own investigations in the field over the last decade, all of which have centred on the conundrum of materialising climate change. If socio-ecological relations are invisible, obscured, or normalised, there is no possibility of effective action. Politicising the systemic degradation of life's ecological foundations is the first and foremost task, and several of our efforts at investigating climate issues have sought to address this.

In 2007, two of us, Rosewarne and Goodman, began researching climate activists involved in direct action protests against fossil fuel industries. The study continued over several years and led to an extended ethnographic reflection on activist experiences, perspectives, and strategies (Rosewarne et al. 2014). At the centre of concern was the need to alert the public to the climate crisis, as the prime symbolic rationale for direct action disruption (and this remains in large part the case). A following project involving most of the current authors was focused on coal mining, asking why new coal mines are approved, and how they are contested. This involved a comparative ethnographic study over four years, focused on three proposed coal mines and mine extensions in India, Germany, and Australia (Goodman et al. 2020). With this we theorised a concept of climate 'articulation', where the structural shifts underway with climate change start to delegitimise coal projects. Anti-coal activists may be motivated by more pressing concerns than climate change, but to a growing degree, their leverage arises from the underlying climate crisis. The contradiction of fossil fuel-based capitalism and climate stability recontextualises and redefines the viability of coal mines, tipping the balance in favour of opponents.

From considering climate and contestation in the transition from coal, we moved to addressing parallel issues in the emergence of renewable energy. A study by two of the present authors, Connor and Heikkinen, involved ethnographic analysis of renewable energy zones in rural New South Wales, Australia, finding a need for greater equity and coordination of the renewable energy transition, to increase benefits and minimise harms for host communities (Cass et al. 2022). In parallel work, Marshall researched barriers to the development of community renewable energy (Buckley et al. 2023; Marshall 2023). Our current project, which we report on in this book, investigates the social legitimacy of renewable energy, with region-level studies, again in India, Germany, and Australia, asking how the decarbonisation of electricity energy is being advanced, and what stands in its way.

Materialising Climate Action

Across these projects, from issues of climate protest to coal phase-out and now to the question of renewables, ethnography has proved to be an invaluable tool to investigate the materialisation of climate, understood as a cultural phenomenon of 'climate worlds' (Hastrup 2015) and as a society-wide process (Baer and Singer 2018). Pressures from the physical impacts of climate change and the climate transition create political dynamics leading to a 'climatisation' of society. We have argued the climate/society gulf is of central importance and must be addressed as a priority (Goodman 2022). The knowledge we have of climate change still dwells principally in the biological and physical sciences, defining the relationship between the carbon cycle and planetary cooling and warming as irresistible 'laws of nature'. These have universal application, making one ton of anthropogenic greenhouse gas emissions notionally 'equivalent' on a global scale, as 'CO_2-e'. The universalising knowledge of CO_2-e stands in stark contrast with the lived experience of energy production and energy transition. There is no universal calculus to 'read-off' the process of emissions reduction. It is always a social process, shaped by the logic of social power and policy as much as by technology and capacity. With this, the clean calculus of CO_2-e can offer cover for the dirty reality of emissions reduction avoidance, in various forms of miscounting, offsetting, and offshoring. Even when successful, emissions reduction at one site can happily be counteracted by expanded emissions elsewhere. Equally, the calculus of CO_2-e can obscure the myriad social dimensions of emissions reduction, not least in terms of who benefits and who loses, aspects that may be critical in determining the prospects for future decarbonisation (Bryant 2019).

This task of 'materialising' the climate crisis is an epistemological challenge as much as a practical one. The clash of knowledges, of scales, and of histories is deeply disruptive of the global social order, not just in terms of impacts but more important in terms of the action required to address it. This may be framed as the intrusion of 'non-human' agency into the dominant capitalist model, but only in the limited sense of a biophysical nexus. The 'agent' in present-day climate change is anthropogenic, and the 'revenge' of nature is only ever a revenge that we wreak on ourselves. In this vein, with this book we aim to stress 'human' agency in the climate nexus and in the wider narrative of the Anthropocene, as it is humans who remake the geology of the non-human planet. By the same token, it is humankind that has the capacity to invoke 'purposeful' climate agency, to restore climate stability, as against 'inadvertent' agency forcing climate instability (Hulme 2010). The epistemological challenge is in this sense both ordinary, in being about human society, and extra-ordinary, in existing far beyond the reach of 'everyday' knowledge and experience. As Amitav Ghosh suggests, with climate change we encounter the 'totality of human actions over time', creating in the present a 'terminus of history' (Ghosh 2023: 167). No surprise we have difficulty addressing it.

As it is unprecedented, climate change is logically unknowable, and hence, in its entirety, beyond experience. As Chakrabarty argues, there is a vast gulf between the narratives of human historical experience and the geological history of climate science. In this context, social agency to address the crisis falls at the first hurdle, as he outlines: 'We may not experience ourselves as a geological agent, but we appear to have become one at the level of the species. And without that knowledge that defies historical understanding there is no making sense of the current crisis that affects us all' (2009: 211). Given this, how may the required climate agency emerge?

Ulrich Beck argues for a method of 'metamorphosis', of holding together both negative and positive dynamics to highlight strategic possibilities, arguing 'it is only when the observer's perspective brings both processes together that new possibilities of action open up' (2015: 78). Climate change is seen as mobilising new publics, bound into a common cause to challenge climate orders. In this respect, it has emancipatory as well as a catastrophic logic, for Beck, an 'emancipatory catastrophism'. Certainly, the cumulative character and resulting fixed horizons of the climate crisis contrast with other sorts of social crisis (see Goodman 2018). Simply put, its irresistible logic forces change. Moore's concept of a 'climate dialectic' (2015), driving an unfolding transformation of political life, is especially useful. Here, the political transformations, and resulting 'purposive' climate agency, arise from the process of contestation over the causes and outcomes of climate change.

In this, the capitalist context is critical: while climate science defines the required measures, it is the regimes of prevailing capitalists that define how they may be (or may not be) achieved. Anthropogenic greenhouse gas emissions are primarily capitalist emissions, driven by the current logic of global-scale accumulation centred on fossil fuels. Addressing decarbonisation therefore means addressing structures of capitalism, and this requires insights from political economy. The analysis of 'greening' capitalism offers a way of apprehending the structural context and its collision with the bio-physical logic of ecological change, in particular with the global greenhouse. The capacity to act on it hinges on agency, which we seek to analyse through ethnographic engagement in specific social situations most affected by the process underway.

Climate Ethnography

In this book we seek a form of normatively engaged 'climate ethnography' to generate transition narratives, both on-the-ground and across differing regional contexts. We use political economy to 'foreshadow' analysis in the sense suggested by Hammersley and Atkinson (2019), to enable strategic selection and interpretation of 'social settings' and ensure that ethnographic inquiry is grounded in the socio-historical and structural context. The ethnography aims in a broad sense to gain an understanding of how people 'make sense' of the world and act in it. Through ethnography we hope to reveal how the social structures emerging under the energy transition constrain and condition action, but also how social agency creates new agendas and transforms those structures. We draw inspiration from the development of ethnographic inquiry and how it has addressed this specific field.

From the 1990s researchers in anthropology became increasingly engaged with climate change and energy transition debates. Initially the focus was on local studies of 'human adaptation to climate change' (Magistro and Roncoli 2001: 95). In 2011 Crate proposed a new category of 'climate ethnography', defined as 'ethnography with a mission' (Crate 2011: 25). Crate was critical of ethnography centred on climate change adaptation, stressing the failure to address the causes of climate change, its social drivers and challengers, or to reveal local-global connections. Crate sought to rearticulate the local focus to make wider claims on climate politics and policy. The refocusing allowed a more dynamic focus on realising climate agency. Peace et al., for instance, argued place-based accounts needed to address the wider problem of how people are 'brought into being' as agents in climate change through contestation (Peace et al. 2012: 33). The extent of change was reflected in the American Anthropological Association's 'Global Climate Change Task Force' report in 2014, which was deeply critical of ethnographic studies that focused only on climate change impacts, 'positioned far more to accommodate change rather than to challenge the causes and drivers' (Fiske et al. 2014: 49). The report called for ethnographers to address solutions, in terms of 'the ethics, practices, and policies/politics of transitioning to a low-carbon society' (Fiske et al. 2014: 72).

The issue of agency and structure is seen as critical. A 2015 collection titled *Grounding Global Climate Change* was focused on using ethnography to bridge climate science and the everyday to produce what the editors called 'social climate research' (Greschke and Tischler 2015). Provocatively, the editors (reflexively) questioned the legitimacy of researchers seeking to bridge divides between climate science and experienced realities, asking, 'does climate change "exist" for an ethnographer if people do not talk about it?' (Greschke and Tischler 2015: 4). The response was both 'yes' and 'no', as the answer depends on context and perspective. Barnes et al., for instance, argued that ethnography can address precisely this disjuncture, providing a 'holistic view' on the interacting levels of abstraction and materiality, where 'at larger spatial and temporal scales, the "fingerprint" of anthropogenic climate change is easy to identify' while at the smaller scale of everyday life, it is 'far more difficult to attribute events and trends to climate change, and project changes and their impacts' (Barnes et al. 2013: 543).

With this, ethnography has opened up new avenues in understanding the social and cultural relations of energy transition. Electricity has been a focus for 'energy ethnography' since the 1940s and is central to 'climate ethnography' (as reflected in the 2010 *Energy Reader*, Nader 2010). There are ethnographies of fossil-fuel dependent communities and how they respond to climate change, for instance, Connor's study of coal in the Hunter Valley in Australia (Connor 2016). And there are ethnographies of energy transition, including the 'Cultures of Energy' collection (Strauss et al. 2013). A special issue of Anthropological Quarterly, edited by Boyer, examined how climate and energy have created new objects of policy and forms of 'energopower' (Boyer 2014). Howe and Boyer's seminal study of the imposition of large-scale corporate-owned wind power on the west coast of Mexico offered a model of critical ethnography orientated to advancing decarbonisation agendas (Howe and Boyer 2019). There is now a growing effort to chart the political dynamics of energy, for instance with the anthology, *Political Anthropology of Energy* (Loloum et al. 2021). There are similar concerns in the field of energy ethics, where energy politics is seen as expressing dilemmas at the heart of society and its ecological relations (High and Smith 2019).

Overall, the normative project of 'climate ethnography' focuses on manifesting and acting on dynamic 'social-ecological relations', rather than simply on mapping the impacts of 'socio-ecological systems' (Goodman 2018). Howe and Boyer's two-volume account of contested wind power in Mexico concludes on the need for government to ensure wide distribution of financial benefit, to involve communities in design and implementation, to maximise broader benefits for the region, and to guarantee transparency and involvement (Howe and Boyer 2019: 194). Normative climate ethnography of this sort can have widespread democratic potential to engage with public and help produce 'purposive' climate agency in the unfolding crisis. From this perspective, climate ethnography can be used to materialise climate change under capitalism and articulate climate social imaginaries, hand-in-hand with climate agency (see Salleh et al. 2016). The approach comes close to Baer and Singer's 'critical political ecology', which seeks to expose social relations of climate change and directly address questions of agency (Baer and Singer 2018). By engaging with the normative horizon, ethnography can help offer the required structural scale and scope to address the capitalocene and its ecological contradictions.

In this book we seek this more holistic approach, setting local contestations in context and arguing that climate change is both produced by socio-ecologies and transforms them. What may seem like a tautology reflects the totality of climate and social relations, where the carbon cycle as a biophysical process is literally remade in the image of capitalist society. In focusing on these issues of contestation and legitimation, we are indebted to Becker and Naumann's ethnographies of energy conflicts in Germany (2018). Our analysis is especially inspired by an important insight that emerges from their work, namely that energy conflicts have a productive or generative dimension. As the legitimacy for transition is challenged, contestation creates new forms of agency and new narratives for legitimation. The local energy conflicts generate social learning, producing new claims for social ownership and benefit-sharing.

To reiterate, normative vision is critical. This entails a focus on dynamics for transformation rather than passive adaptation. Unmitigated climate change does not allow for an adapted future and requires strategic contestation across multiple social fields. Contestation makes visible the strategic possibilities and makes them accessible and necessary.

Contrast Cases

This study embraces aspects of interdisciplinarity, narration, normative intent, and articulation. It deploys these through a comparison of three cases to draw common lessons. With this it adopts a 'contrast cases' approach, where commonalities are sought across radically contrasting contexts. The comparative case study method is driven by wider research questions to illustrate wider themes. Comparative case analysis can then aim to explain similarities and differences, for instance in historical sociology, to learn from the past in order to gain insights for the present

(George and Bennett 2005; Lange 2013). In this study we seek lessons from located experiences of renewable energy developments, to shed light on the decarbonisation of electricity: the comparative approach thus engages normative intent in aiming to generate specific policy parameters to promote renewable energy, as well as to understand the process that it puts into play and where it may be leading. We undertake comparative analysis 'to make sense of differences, not to collapse them' (Strathern 1987: 286); in Hastrup's terms, we ask 'how we can generate genuinely new, general knowledge on the basis of particularities' (2015: 139–140).

These insights are sought across three contrasting countries: India, Germany, and Australia. India is a rapidly industrialising country, with rising electricity demand, fed mainly by coal. Germany is post-industrial, though it still mines brown coal, while exporting renewable energy to its neighbours. Australia is extractivist, advancing domestic renewable energy while remaining one of the world's largest fossil fuel exporters. All three countries are avid champions of renewables and, to a varying degree, embrace energy transition. Within each country we focus on regions that have become especially dependent on renewable energy for electricity, namely Karnataka in India, Brandenburg in Germany, and the Upper Spencer Gulf in South Australia.

In each of these regions, we delve into the interplay of local contexts and social action to illuminate how renewable energy initiatives are conceived, implemented, and experienced on the ground. With this we aim to understand the emerging socio-ecological 'fix' as it unfolds, exploring how communities navigate the shift from fossil fuel dependency to renewable energy. Ethnography allows for an understanding of local responses, social appropriations, and the emergence of new social claims, uncovering the nuances of lived realities, knowledge worlds, conflicts, and opportunities embedded in the transition. Through this we gain an in-depth understanding of located experiences, perspectives, and behaviours, illustrating power relations and conflicts embedded in the transition, uncovering ways in which renewable energy practices intersect with existing social structures, including issues of ownership, work structures, and community well-being. Overall, by merging the macro-level political economy with more micro-level ethnographic insights, we aim to highlight the barriers, possibilities, and lessons for energy transitions across the markedly different socio-political contexts and thereby draw wider lessons for rapid energy transition and emissions reduction elsewhere.

Conclusions

The intent of this opening chapter has been to debate a range of perspectives on the emerging renewable energy socio-ecological transformation. We have sought to provide some theoretical depth to the idea of socio-ecological relations and how it may be deployed to interpret historical energy transitions. We then used the concept to ground analysis of the possibilities and barriers to renewable energy, distinguishing the three dimensions of accumulation, appropriation, and regulation. First, we emphasise the accumulation drive for surplus, examining how it manifests in the 'business models' governing energy production, profit, and growth. Second, we identify the unique logic of ecological appropriation within energy production, exploring its implications for social organisation. Third, we address the challenges of state-led regulation and rivalry in the absence of collectively agreed foresight or an agreed agenda. The three dimensions are intertwined and have a contradictory logic, with accumulation often occurring at the expense of subordinated classes, appropriation benefiting elites while degrading ecologies, and inter-state rivalry facilitating domination over both people and nature.

Moving forward, Chapter 2 shifts the focus to narratives of renewable energy regionalism, analysing the political economy of transition in terms of legitimation and capitalisation within each of the three case studies. Much existing analysis has concentrated on the influence of the national state, and this chapter seeks to focus attention on regions and their significance in the shift from fossil-fuel-based electricity generation, particularly as large-scale wind and solar farms emerge predominantly in regional and remote areas. This contextual analysis of the three cases sets the stage for the subsequent case studies in Chapters 3–5, which develop ethnographic accounts to explore how the social-ecological relations of renewable energy unfold on the ground. In each of the case study chapters, the focus is on the emergence of renewable energy narratives and how they have developed over time at the regional level. The concluding comparative chapter, Chapter 6, seeks to draw together lessons from the three cases, returning to the conceptual schema established in this chapter. Across these chapters, it is important to emphasise the uncertainties and contested nature of the renewable energy transition. Each of the dimensions of energy transition is patterned by relations of dominance and subordination and offers the potential for contestation and change as part of the socio-ecological relations of renewable energy expansion.

References

Albert, M. (2022). COVID-19 and the planetary crisis multiplicity: from Marxist crisis theory to planetary assemblage theory. *Theory & Event*, 25(2), 332–363.

Angus, I. (2023). The meaning of so-called primitive accumulation. *Monthly Review*, 74(11), 1–10.

Araújo, K. (2014). The emerging field of energy transitions. *Energy Research & Social Science*, 1, 112–121.

Armstrong, J. (2021). People and power: expanding the role and scale of public engagement in energy transitions. *Energy Research & Social Science*, 78, 102136.

Backhouse, M. & Lehman, R. (2020). New 'renewable' frontiers: contested palm oil plantations and wind energy projects in Brazil and Mexico. *Journal of Land Use Science*, 15(2), 373–388.

Baer, H. & Singer, M. (2018). *The Anthropology of Climate Change: An Integrated Critical Perspective*, London: Routledge.

Bainton, N. & Kemp, D. (2021). 'More clean energy means more mines – we shouldn't sacrifice communities in the name of climate action'. *The Conversation*, 4 November 2021.

Baker, L. (2022). Procurement, finance and the energy transition: between global processes and territorial realities. *Environment and Planning E: Nature and Space*, 5(4), 1738–1764.

Barnes, J. et al. (2013). Contribution of anthropology to the study of climate change. *Nature Climate Change*, 3(6), 541–544.

Barry, J. (2021). Green republicanism and a 'just transition' from the tyranny of economic growth. *Critical Review of International Social and Political Philosophy*, 24(5), 725–742.

Beck, U. (2015). Emancipatory catastrophism: what does it mean to climate change and risk society? *Current Sociology*, 63(1), 75–88.

Becker, S. & Naumann, M. (2018). Energiekonflikte erkennen und nutzen (Identifying and Using Energy Conflicts). In O. Kühne & F. Weber, eds., *Bausteine der Energiewende (Building-Blocks of the Energy Transition)*, Wiesbaden: Springer VS, pp. 509–522.

Bond, P. (2020). Climate justice advocacy: strategic choices for Glasgow and beyond. In S. Bohm & S. Sullivan, eds., *Negotiating Climate Change in Crisis*, Cambridge: Open Book Publishers, pp. 335–354.

Boyer, D. (2014). Energopower. *Anthropological Quarterly*, 87(2), 309–333.

Bridge, G. et al. (2020). Pluralizing and problematizing carbon finance. *Progress in Human Geography*, 44(4), 724–742.

Bridge, G. & Gailing, L. (2020). New energy spaces: towards a geographical political economy of energy transition. *Environment and Planning A: Economy and Space*, 52(6), 1037–1050.

Bruckner, T., et al. (2014). Energy systems. In O. Edenhofer, et al., *Climate Change 2014: Mitigation of Climate Change. Contribution of Working Group III to the Fifth Assessment Report of the Intergovernmental Panel on Climate Change*, Cambridge: Cambridge University Press, pp. 511–598.

Bryant, G. (2019). *Carbon Markets in a Climate-Changing Capitalism*, Cambridge: Cambridge University Press.

Buckley, E., Walters, K., Marshall, J. & Ford, A. (2023). *Australian Community Energy Collective Impact Assessment*, Sydney: Community Power Agency.

Buller, A. (2022). *The Value of a Whale: On the Illusions of Green Capitalism*, Manchester: Manchester University Press.

Burkett, P. & Foster, J. (2006). Metabolism, energy, and entropy in Marx's critique of political economy. *Theory and Society*, 35(1), 109–156.

Carroll, W. (2020). Fossil capitalism, climate capitalism, energy democracy: the struggle for hegemony in an era of climate crisis. *Socialist Studies*, 14(1), 1–28.

Cass, D., Connor, L., Heikkinen, R., & Pearse, R. (2022). *Renewables & Rural Australia*, Sydney: Sydney Environment Institute and The Australia Institute.

Chakrabarty, D. (2009). The climate of history: Four theses. *Critical Inquiry*, 35(2), 197–222.

Chakrabarty, D. (2022). *A Climate of History in a Planetary Age*, Chicago: Chicago University Press.

Christophers, B. (2024). *The Price Is Wrong: Why Capitalism Won't Save the Planet*, London: Verso.

Connor, L. (2016). *Climate Change and Anthropos: Planet, People and Places*, London: Routledge.

Crate, S. (2011). Climate and culture: anthropology in the era of contemporary climate change. *Annual Review of Anthropology*, 40(1), 175–194.

Dunlap, A. (2021). End the 'green' delusions: industrial-scale renewable energy is fossil fuel+. In S. Bohm & S. Sullivan, eds., *Negotiating Climate Change in Crisis*, Cambridge: Open Book Publishers, pp. 113–126.

Ellis, G., Schneider, N. & Wüstenhagen, R. (2023). Dynamics of social acceptance of renewable energy: an introduction to the concept. *Energy Policy*, 181, 113706.

Engels, F. (1953). *The Part Played by Labour in the Transition from Ape to Man*, first published 1876, Moscow: Progress Publishers.

Eriksen, T. (2016). *Overheating: An Anthropology of Accelerated Change*, London: Pluto Press.

Fast, S. (2013). Social acceptance of renewable energy: trends, concepts, and geographies. *Geography Compass*, 7(12), 853–866.

Featherstone, D. (2011). On assemblage and articulation. *Area*, 43(2), 139–142.

Finley-Brook, M. & Holloman, E. (2016). Empowering energy justice. *Environmental Research and Public Health*, 13(9), 926.

Fiske, S., Crate, S., Crumley, C., Galvin, K., Lazrus, H., Luber, G., Lucero, L., Oliver Smith, A., Orlove, B., Strauss, S. & Wolk, R. (2014). *Final Report of the AAA Global Climate Change Task Force*, Arlington, US: American Anthropological Association.

Ford, A. & Newell, P. (2021). Regime resistance and accommodation: toward a neo-Gramscian perspective on energy transitions. *Energy Research and Social Science*, 79, 102163.

Foster, J. (2009). *The Ecological Revolution*, New York: Monthly Review Press.

Freese, B. (2003). *Coal: A Human History*, New York: Perseus Books.

Gazmararian, A. & Tingley, D. (2023). *Uncertain Futures: How to Unlock the Climate Impasse*, Cambridge: Cambridge University Press.

George, A. & Bennett, A. (2005). *Case Studies and Theory Development in the Social Sciences*, Cambridge: MIT Press.

Ghoddousi, P. & Page, S. (2020). Using ethnography and assemblage theory in political geography. *Geography Compass*, 14(10), 12533.

Ghosh, A. (2023). *The Great Derangement: Climate Change and the Unthinkable*, Chicago: University of Chicago Press.

Gills, S. & Morgan, J. (2021). Postscript, an end to the war on nature: COP in or COP out? *Globalizations*, 18(7), 1311–1322.

Goldman, M. (1998). *Privatising Nature: Political Struggles for the Global Commons*, London: Pluto Press.

Goodman, J. (2018). Researching climate crisis and energy transitions: some issues for ethnography. *Energy Research & Social Science*, 45, 340–347.

Goodman, J. (2022). Social movements and climate change: climatizing society from within. In H. Von Storch, et al., eds., *Oxford Research Encyclopedia of Climate Science*, Oxford: Oxford University Press, pp. 1–18.

Goodman, J. et al. (2020). *Beyond the Coal Rush: Turning Point for Global Energy and Climate Policy?*, New York: Cambridge University Press.

Greschke, H. & Tischler, J. (2015). *Grounding Global Climate Change: Contributions from the Social and Cultural Sciences*, Dordrecht: Springer.

Gross, C. (2007). Community perspectives of wind energy in Australia. *Energy Policy*, 35(5), 2727–2736.

Haberi, H. et al. (2020). A systematic review of the evidence on decoupling of GDP, resource use and GHG emissions: synthesizing the insights. *Environmental Research Letters*, 15(6), 065003.

Hammersley, M. & Atkinson, P. (2019). *Ethnography: Principles in Practice*, 4th edition, London: Routledge.

Haney, A. & Pollitt, M. (2013). New models of public ownership in energy. *International Review of Applied Economics*, 27(2), 174–192.

Hastrup, K. (2015). Comparing climate worlds: theorising across ethnographic fields. In H. Greschke & J. Tischler, eds., *Grounding Global Climate Change*, Dordrecht: Springer, pp. 139–154.

Helfer, F. (2022). Using assemblage theory in political anthropology. *Hamburger Journal für Kulturanthropologie*, 14, 33–47.

High, M. & Smith, J. (2019). Energy ethics. special issue, *Journal of The Royal Anthropological Institute*, 25(1), 9–28.

Hoffman, J., Davies, M., Bauwens, T., Späth, P., Hajer, M., Arifi, B., Bazaz, A. & Swilling, M. (2021). Working to align energy transitions and social equity: an integrative framework linking institutional work, imaginaries and energy justice. *Energy Research & Social Science*, 82, 102317.

Howe, C. & Boyer, D. (2019). *Wind and Power in the Anthropocene*, Durham: Duke UP.

Huber, M. (2022). *Climate Change as Class War*, London: Verso.

Hughes, T. (2004). *Human-Built World: How to Think about Technology and Culture*, Chicago: Chicago University Press.

Hulme, M. (2010). Mapping climate knowledge: an editorial essay. *Wiley Interdisciplinary Reviews: Climate Change*, 1(1), 1–8.

Jungjohann, A. & Morris, C. (2016). *Energy Democracy. Germany's Energiewende to Renewables*, Wiesbaden: Springer.

Kern, F. & Rogge, K. (2016). The pace of governed energy transitions: agency, international dynamics and the global Paris agreement accelerating decarbonisation processes? *Energy Research and Social Science*, 22, 13–17.

Knuth, S., Behrsin, I., Levenda, A. & McCarthy, J. (2022). New political ecologies of renewable energy. *Environment and Planning: Nature and Space*, 5(3), 997–1013.

Lange, M. (2013). *Comparative-Historical Methods*, London: Sage.

Laxer, G. & Soron D. (2006). *Not for Sale: Decommodifying Public Life*, Toronto: University of Toronto Press.

Lecocq, F., et al. (2022). Mitigation and development pathways in the near- to mid-term. In P. Shukla, et al., eds., *Climate Change 2022: Mitigation of Climate Change. Contribution of Working Group III to the Sixth Assessment Report of the Intergovernmental Panel on Climate Change*, Cambridge: IPCC and Cambridge University Press, pp. 202–409.

Leonard, A., Ahsan, A., Charbonnier, F. & Hirmer, S. (2022). The resource curse in renewable energy: a framework for risk assessment. *Energy Strategy Reviews*, 41, 100841.

Loloum, T., Abram, S. & Ortar, N. (2021). *Ethnographies of Power. A Political Anthropology of Energy*, New York: Berghahn Books.

Magistro, J. & Roncoli, C. (2001). Anthropological perspectives and policy implications of climate change research. *Climate Research*, 19(2), 91–96.

Malm, A. (2016). *Fossil Capital: The Rise of Steam Power and the Roots of Global Warming*, London: Verso.

Marcus, G. & Saka, E. (2006). Assemblage. *Theory, Culture & Society*, 23(2), 101–106.

Marshall, J. (2023). Comparing local energy conflicts in NSW Australia: moving to climate generosity. *Globalizations*, 20(8), 1363–1379.

Marx, K. (1976). *Capital: A Critique of Political Economy, Volume 1*, New York: Penguin.

McCarthy, J. (2015). A socioecological fix to capitalist crisis and climate change? The possibilities and limits of renewable energy. *Environment and Planning A*, 47(12), 2485–2502.

Michalena, E. & Hills, J. M. (2013). *Renewable Energy Governance' Complexities and Challenges*, London: Springer.

Mitchell, T. (2013). *Carbon Democracy: Political Power in the Age of Oil*, London: Verso.

Mookerjea, S. (2019). Renewable energy transition under multiple colonialisms: passive revolution, fascism redux and utopian praxes. *Cultural Studies*, 33(3), 570–593.

Moore, J. (2015). *Capitalism in the Web of Life: Ecology and the Accumulation of Capital*, London: Verso.

Morton, A. (2010). The continuum of passive revolution. *Capital and Class*, 34(3), 315–342.

Nader, L. (2010). *The Energy Reader*, New York: Wiley Blackwell.

Nail, T. (2017). What is an assemblage? *SubStance*, 46(1), 21–37.

Newell, P (2021). *Power Shift: The Global Political Economy of Energy Transitions*, Cambridge: Cambridge University Press.

O'Connor, J. (1998). *Natural Causes: Essays in Ecological Marxism*, New York: Guilford Press.

Paterson, M. (2010). Legitimation and accumulation in climate change governance. *New Political Economy*, 15(3), 345–368.

Peace, A., Connor, L., & Trigger, D. (2012). Environmentalism, culture, ethnography. *Oceania*, 82(3), 217–227.

Pearse, R., & Bryant, G. (2022). Labour in transition: a value-theoretical approach to renewable energy labour. *Environment and Planning E: Nature and Space*, 5(4), 1872–1894.

Polanyi, K. (1944). *The Great Transformation: The Political Origins of Our Time*, Boston: Beacon Press.

Radtke, J. & Scherhaufer, P. (2022). A social science perspective on conflicts in the energy transition: an introduction to the special issue. *Utilities Policy*, 78, 101396.

Rentschler, J. & Bazilian, M. (2017). Reforming fossil fuel subsidies: drivers, barriers and the state of progress. *Climate Policy*, 17(7), 891–914.

Reusswig, F., Braun, F., Hegera, I., Ludewig, T., Eichenauer, E. & Lassa, W. (2016). Against the wind: local opposition to the German 'Energiewende'. *Utilities Policy*, 41, 214–227.

Ribiero, F., Ferreira, P. & Araujo, M. (2011). The inclusion of social aspects in power planning. *Renewable and Sustainable Energy Reviews*, 15(9), 4361–4369.

Rockström, J. et al. (2024). The planetary commons: a new paradigm for safeguarding Earth-regulating systems in the Anthropocene. *PNAS*, 121(5), 2301531121.

Rosewarne, S. (2022). *Contested Energy Futures: Capturing the Renewable Energy Surge in Australia*, Singapore: Palgrave Macmillan.

Rosewarne, S., Goodman, J. & Pearse, R. (2014). *Climate Upsurge: The Ethnography of Climate Movement Politics*, London: Routledge.

Routledge, P., Cumbers, A. & Derickson, K. (2018). States of just transition: realizing climate justice through and against the state. *Geoforum*, 88, 78–86.

Saito, K. (2022). *Marx in the Anthropocene: Towards the Idea of Degrowth Communism*, Cambridge: Cambridge University Press.

Salleh, A., Goodman, J. & Hosseini, H. (2016). From sociological to ecological imagination. In J. Marshall & L. Connor, eds., *Environmental Change and the World's Futures: Ecologies, Ontologies, Mythologies*, London: Routledge, pp. 96–109.

Sareen, S. & Haarstad, H. (2020). Legitimacy and accountability in the governance of sustainable energy transitions, *Global Transitions*, 2, 47–50.

Shukla, P., Skea, J. & Reisinger, A. (2022). Summary for policymakers. In *Climate Change 2022: Mitigation of Climate Change. Contribution of Working Group III to the Sixth Assessment Report of the Intergovernmental Panel on Climate Change*, IPCC and Cambridge: Cambridge University Press, 3–48.

Sinn, H. (2012). *The Green Paradox: A Supply-Side Approach to Global Warming*, Cambridge US: MIT Press.

Smith, J. & High, M. (2017). Exploring the anthropology of energy: ethnography, energy and ethics. *Energy Research and Social Science*, 30, 1–6.

Smith, N. (1984). *Uneven Development: Nature, Capital and the Production of Space*, Oxford: Blackwell.

Sovacool, B., Burke, M., Baker, L., Kotikalapudi, C. & Wlokas, H. (2017). New frontiers and conceptual frameworks for energy justice. *Energy Policy*, 105, 677–691.

Strathern, M. (1987). An awkward relationship: The case of feminism and anthropology. *Signs* 12(2), 276–292.

Steffen, B., Karplus, V. & Schmidt, T. (2022). State ownership and technology adoption: the case of electric utilities and renewable energy. *Research Policy*, 51(6), 104534.

Strauss, S., Rupp, S. & Love, T. (2013). *Cultures of Energy: Power, Practices, Technologies*, London: Routledge.

Stuart, D., Gunderson, R. & Peterson, B. (2020). The climate crisis as a catalyst for emancipatory transformation: an examination of the possible. *International Sociology*, 35(4), 433–456.

Szulecki, K. & Overland, I. (2020). Energy democracy as a process, an outcome and a goal: a conceptual review. *Energy Research and Social Science*, 69, 101768.

Temper, L. et al. (2020). Movements shaping climate futures: a systematic mapping of protests against fossil fuel and low-carbon energy projects. *Environmental Research Letters*, 15, 123004.

Tokar, B. & Gilbertson, T. (2020). *Climate Justice and Community Renewal: Resistance and Grassroots Solutions*, London: Routledge.

Thomson, I. & Boutilier, R. (2011). Social license to operate. In P. Darling, ed., *Mining Engineering Handbook*, Littleton: Society for Mining, Metallurgy and Exploration, pp. 1779–1796.

Treu, N., Schmelzer, M. & Buckhart, C. (2020). *Degrowth in Movement(s): Exploring Pathways for Transformation*, Winchester: John Hunt Books.

UNEP (2023). *Emissions Gap Report 2023*, New York: UNEP.

UNFCCC (2015a). *Paris Agreement*, New York: UN.

UNFCCC (2015b). *Synthesis Report on the Aggregate Effect of the Intended Nationally Determined Contributions*, New York: UN.

UNFCCC (2023). 'First Global Stocktake, Conference of the Parties serving as the meeting of the Parties to the Paris Agreement', United Nations, 13 December 2023.

Unruh, G. (2006). Globalizing carbon lock-in. *Energy Policy*, 34(10), 1185–1197.

Urry, J. (2014). The problem of energy. *Theory, Culture and Society*, 31(5), 3–20.

Wahlund, M. & Palm, J. (2022). The role of energy democracy and energy citizenship for participatory energy transitions: a comprehensive review. *Energy Research & Social Science*, 87, 102482.

Wood T. & Dundas, G. (2020). *Start with Steel: A Practical Plan to Support Carbon Workers and Cut Emissions*, Melbourne: Grattan Institute.

Wüstenhagen, R., Wolsink, M. & Bürer, M. (2007). Social acceptance of renewable energy innovation: an introduction to the concept. *Energy Policy*, 35(5), 2683–2691.

Wolsink, M. (2012). The research agenda on social acceptance of distributed generation in smart grids: renewable as common pool resources. *Renewable and Sustainable Energy Reviews*, 16(1), 822–835.

2

Renewable Energy Regions

Cases and Policy

Sub-national regions are becoming key drivers of energy transition. National governments often submit targets for expanded renewable energy to the United Nations as part of their Intended Nationally-Determined Contributions (INDCs) for emissions reduction. These targets have their most salient material impact on emerging renewable energy regions and on regions dependent on fossil fuel industries. National-level commitments are critical to the transition, but renewable energy installations are place-based and require the active involvement of sub-national authorities. The regional context can be framed as a microcosm, revealing how the wider pressures shape outcomes. Localities can be recast as symbolic of the wider societal transformations, and vice versa, transposing the national-level transition into locality-centred contexts.

The lynchpin of emissions reduction is the installation of renewable energy-generating capacity, and this requires stable access to land (or sea) and the capacity to create new energy geographies. With this, national-to-sub-national relations and inter-regional and intra-regional dynamics become important factors in renewable energy development. Regional energy governance becomes central in terms of the role of energy corporations, whether fossil fuel or renewable, as well as climate and energy organisations, both governmental and non-governmental, and the political organisation of regional citizen-voters. Emphasising these institutional factors, for instance, Bridge and Gailing (2020) develop a region-centred model extending across spatial unevenness, regional agglomeration, and energy governance strategies.

In this book we focus on regions that have become heavily reliant on renewable energy, investigating how they achieved this, what legitimacy challenges they have faced, and what this suggests about future trajectories. The first part of this chapter discusses the regional character of the transition to renewable energy; the second part provides an historical introduction to each of the three cases, prefacing what is to follow in Chapters 3 to 5.

Conceptualising Renewable Energy Regions

There is a growing interest in how to theorise the role of regions in the transition from fossil-fuel electricity generation to renewable sources of energy, especially in the establishment of large-scale wind and solar farms. The term 'energy regionalism' has emerged at the international level, as inter-state 'macro-regions' collaborate and compete for access to energy resources (Hancock 2024). Our focus here is on 'micro-regions' as sub-national units that emerge as key sites for renewable energy reflecting a new sub-national energy geography. Most studies of energy transition have focused on national-level contexts (Jenniches 2018), but there are efforts to address the dynamics at the region level, such as a special issue of *Economy and Space* on 'New Energy Spaces' (Bridge and Gailing 2020). Bridge and Galling's introduction to that special issue shows that the geographical potentials of emergence and transformation are now important policy issues. There is a growing focus on transitions as making new and highly uneven energy geographies, as an instance of space-making in capitalist geography. These approaches seek to draw together traditions in local and regional geography that focus on the making of sub-national geographies, especially in terms of regional policy and strategy. Analysis of 'post-Fordist' geographies in the 1980s addressed the regionalisation of industrial organisation under neoliberalism. Debates about regional agglomeration as a governance strategy overlapped with the growing regional unevenness between winners and losers under neoliberalism, begging the question of more effective regional strategy (Burroni 2014; Mattes et al. 2015).

Echoing studies of industrial regionalism, regional bodies are increasingly recognised as leading players in renewable energy development. Early research highlighted the importance of regional 'guiding visions' in energy transitions, in integrating various dimensions and playing a convening, institutionalising, and stabilising role (Späth 2010). The importance of the regional state is emphasised, along with the governance frameworks to maximise local spillover effects, distribute benefit, and underpin legitimation. This emergence of renewable energy regionalism is interpreted as a category in the making, and open for development, including posing the possibility of a regional renewable energy 'commons' (Moss, Becker and Naumann 2015; Johnson and VanDeer 2021).

Across these debates, regional governance is seen as central. De Laurentis and Pearson (2021) suggest three aspects shape outcomes: renewable energy sources and land, policy vision and narrative, regulation, and infrastructures. They find that national-level renewables policy can cause friction: 'There is a need for regional energy strategies and policies to be closely aligned with land-use planning processes and institutions … balancing trade-offs between multiple energy and non-energy objectives' and making synergies 'salient at the region level' (De Laurentis and Pearson 2021: 16).

Regional governance and planning can thereby redefine regional identity in terms of the benefit from renewables and vice versa (Mattes et al. 2015). Partly reflecting this, region-level governance for renewable energy development appears to be strengthening: in Germany, for instance, there are 233 regional energy plans that have been created (Lutz et al. 2017). Hoberg argues that contestation of renewables primarily takes the form of 'place-based resistance' against energy installations, and that this begs the question of regional capacity. He stresses a national-local calculus, where regions are positioned as sacrificing for national benefit. Cost and benefit are then defined and politicised spatially, 'between local desires to minimise impacts and the broader public interest in establishing necessary infrastructure' (Hoberg 2021: 298). Regional governance then becomes vitally important in terms of enabling alternative pathways for otherwise neglected regions.

As we have outlined, configurations of ecology, land, technology, policy, and profit as expressed and vested in the new renewable energy installations deliver a new 'fix' for investment capital. Renewable energy is dependent on effective regional policies to facilitate the 'harvesting' of wind and solar power (Elavarasan [2020] explicitly uses the term). A new renewable energy geography emerges in 'greenfield' regions that often have no history of mining, power generation, or heavy industry. Renewable energy can also have major impacts in post-industrial 'brown fields', where skilled labour supply and infrastructure, especially electricity transmission, can be regeared for renewables. Across these contexts, the regulatory regime is constructed in anticipation of transition, correlating land, finance, and institutional power for the new forms of regional development, redefining regional 'identity'.

The focus of policy across the cases examined for this book is on attracting private sector investment: the structures for renewable energy governance facilitate investment in wind and solar parks and private installers engage with regional communities, who may embrace or contest the emerging industries. New renewable energy planning zones emerge and the local state collaborates with landholders who see financial advantage in hosting renewable energy generation and transmission. In each of the case studies examined here, the resulting renewable energy policy is founded on assumptions about the virtues of private sector involvement and the facilitative purpose of state authority, to underpin corporate interests in the renewable sector. Yet, as we document, the state is the central player: the transition to renewable energy and the remaking of energy geography cannot proceed without the exercise of state agency.

National State Policy and Regions

As a rule, the national state seeks to ensure reliability and affordability of energy supply and controls supply frequency and voltage volume to achieve this. In contrast with hierarchically structured electricity, the process of corporatising and privatising state-run energy creates access for independent generators to the centralised grid. Governments have established managed 'energy markets', where electricity pricing becomes highly volatile. The emergence of diverse private electricity generators, from privatised utilities to a growing cohort of household and small-scale generators, all simultaneously feeding into the grid, makes system stability harder to manage. In this context the state's reliance on private capital and electricity markets to deliver decarbonisation creates greater risks, limitations, and dependencies: capital flows and markets are notoriously unreliable and subject to boom and bust, and only have a fixed capacity.

States have long sought to underpin energy supplies, especially electricity supply, as a key aspect of economic development: electricity supply and transmission projects are often defined as national priority projects to remove planning impediments. Under neoliberalism, the state has remained central to the electricity sector, but, as noted, its role has become more geared to facilitating private finance and less to securing public provision. The institutional and organisational transformations wrought by the construction of energy 'markets' and the privatisation of state-owned power utilities have shaped the way in which the transition to renewable energy is implemented. This in turn has had an impact on the forms of new energy geography that are emerging. As we discuss in the case studies, among the more important elements has been the decentralisation of electricity generation via 'offtake', or as termed in Australia, 'Power Purchase Agreements' (PPAs). PPAs bypass wholesale markets and lock in the long-term price of energy contracts with state agencies and other organisations. They thereby provide a de-risking device that enables investment in region-based solar and wind farms, while also defining follow-on requirements for extended and reconfigured transmission networks.

As new sites of accumulation, global renewable energy corporations and international investment funds dedicated to decarbonisation can exert a significant influence on new energy regions and have played an important role in advancing the transition from the coal-fired emissions-intensive system. Energy 'markets' and reverse auctions for PPAs encourage price competition but also entrench the larger corporate players that are more able to compete. With growing price volatility, state authorities have enabled transactions in hedging, electricity futures, and other derivatives, which can be traded against electricity wholesale and spot prices. This has produced a deeper financialisation of energy, especially renewable energy, which again entrenches the larger players and those more able to leverage diverse holdings and feeds energy speculation with resulting destabilising effects.

At one level, the proportion of renewable generation in the energy mix increases by attracting the ensuing capital flow; at another level, the process of decarbonisation, and climate stability more broadly, is made hostage to rates of return for the large investor and to the market instabilities exacerbated by energy speculation. With this, transition is also hostage to ever-expanded energy use, despite the necessity as outlined earlier, for reduced energy use in order to meet the required emissions reduction targets. Reflecting this, there is a need to articulate ecological modernisation and degrowth perspectives in debates about regional energy transitions (Gibbs and O'Neil 2017).

Regional and Inter-regional Pressures

With national bodies seeking to manage electricity markets, the process of enabling the transition to renewables falls to sub-national authorities. The sub-national state gains responsibility and capacity to pursue region-level development strategies and to envisage new economic opportunities in a putative renewable economy. As noted, state authorities can become highly catalytic in producing new modes of renewable developmentalism, for instance, enabling access to the land and infrastructure necessary for wind and solar farms and facilitating both 'upstream' and 'downstream' linkages into the regional economy. With this, the sub-national state becomes a key site of governance and agency for energy transition, thereby redefining the socio-ecological relations of the region and its energy scape.

A country's 'energy mix' is thus in practice a collection of regional 'energy mixes'. Regional unevenness shapes the outcomes. In all three counties, India, Germany, and Australia, some regions have developed a focus on heavy industry linked with coal-fired power and local coal mining. In India, there is also dependence on imported coal, whether internationally or within the national context. Sub-national fossil fuel dependence may produce a close relationship between the local state and fossil fuel interests, undermining renewables; the converse may apply for regions with minimal fossil fuel availability, where strong regional linkages can emerge with the renewables sector. If renewable energy is concentrated in regions without access to local fossil fuels, inter-regional energy divergence can increase, such as occurred in Australia, between South Australia and coal-rich states such as New South Wales and Queensland; in India, eight of thirty-six states and territories have wind and solar dependence at more than double the national average, including Karnataka at almost quadruple (IEA 2021). Renewables regions then develop different intra-regional industrial social and spatial structures, characterised, for instance, by utility-scale wind and solar in rural areas, concentrated around transmission access, as against pre-existing fossil fuel generation located in close proximity to coal fields or to import terminals.

These region-level developments are set in the context of wider pressures on energy policy. Climate disruption and global climate policy set the overarching frame for energy transition, which then can be reinforced by other imperatives. For instance, in the context of Russia's invasion of Ukraine, energy security has become a primary concern in the European Union, expressed in the 2022 'REpowerEU' programme, which prioritises renewable energy to replace imported Russian gas supplies (European Commission 2022). At the same time, there is continued and in some respects expanded extraction of coal and gas, which can exacerbate regional divergence. This is particularly marked in Australia where most coal and gas is exported, and is rapidly expanding with rising fossil fuel prices (again following the Ukraine invasion). The countervailing pressures reflect the logic of transition as a state of flux between fossil fuels and renewables and arguably are being overcome as coal and gas 'phases-down' for domestic power generation across the three cases. Regional agglomerations of vested interests and entrenched power can be the deciding factor determining whether renewable energy regions, and consequently the low-carbon energy transitions, are slowed or hastened (Späth and Rohracher 2010). Such agglomerations hinge on socio-political factors as much as on administrative or technical capacity.

Regions do not exist in isolation: sub-national authorities both learn and compete with each other to make gains in the emerging regional renewable economy. The multilevel character of energy policy, reflecting the particular history of state formation and the process of electricity generation and distribution, can offer surprising autonomy for regional bodies. Here the exercise of regulatory power for a national approach to energy development can come into direct conflict with sub-national efforts at achieving regional energy transition. The vertical jurisdictional tension then is deployed politically, either from 'above' or from 'below'. Horizontal pressures can be equally important in terms of inter-regional dynamics, which can be rivalrous as well as collaborative. Divergence in the politics of renewables between national and regional levels, or between regions, may hamper renewables development, although jurisdictional rivalry may equally spur regional mobilisation and innovation in renewable governance.

The intra-regional aspects of this process can be critical as renewable energy is concentrated in particular localities, affecting some sub-regions and not others. There are also powerful multi-scalar alignments, not least with large global investors and corporates drawn to the local renewable planning zones, mediated through the layers of state policy. The multi-scalar process can itself become rivalrous, with localities or regions seeking to attract footloose energy capital or seeking to compete for favour in the process of national policymaking. There is strong evidence that institutional coherence across the levels of authority, local to national, aids in the development of renewable energy. A study found Germany significantly more effective due to a multilevel consensus about the purpose of energy policy under climate change, in contrast with Australia (Jehling et al. 2019). There are parallel findings in other federal contexts, from the United States (Stokes 2022) to Russia (Boute 2013). Overall, regional autonomy and variation can be generative, with new modes of social organisation along with technical applications, linked with specific local spatial fixes, driving a process of experimentation. The net result is growing regional unevenness as some regions become increasingly integrated into flows of international investment finance, with regional nodes defined against hinterlands and vice versa.

Regionalisation in the Case Studies

India, Australia, and Germany are federal states, each with a history of sub-national regional autonomy in energy policy. They contrast in terms of their history of state formation and their current developmental trajectory and also in terms of the national energy mix. India is most dependent on fossil fuels for generation (with 74% of energy reliant on coal and 5.9% natural gas) (Chye and Chew 2023). Coal accounts for 49% of energy generation in Australia and gas 19%, while total fossil fuel reliance stands at 68% (DCCEWE 2023); Germany is the least dependent on hard coal and lignite, which are used to generate 35.3% of Germany's power, with natural gas and nuclear accounting for 12.8 and 11.7%, respectively (Statistisches Bundesamt 2022; Appunn 2023).

The reliance on renewable energy sources is increasing overall, although as noted there is considerable variation at the sub-state regional level. In India, the divergence is presenting many states with 'integration' challenges, with growing excess daytime energy in states with higher dependence on renewable energy (IEA 2021). As noted, the share of solar and wind in the electricity provision was highest in in India's renewables-rich states – Gujurat, Rajastan, Karnataka, Himachal Pradesh, and Tamil Nadu. In Karnataka, solar and wind energy accounted for about 55% of installed capacity and 43% of the state's electricity generation in 2023 (MNRE 2024b). In Australia (eight states and territories), both South Australia and Tasmania have a very high reliance on renewables. In South Australia, wind and solar account for 71% of electricity, with wind at 44% and solar at 26%. Virtually all electricity in Tasmania is renewable, at 98%, but this is mostly hydroelectricity, which produces 80% (DCCEWE 2023). In Germany, different meteorological and topographical conditions are reflected in diverse renewable energy outcomes across the country's 16 states or *Länder* (Pinkera 2020). Marked variations in setting transition targets have been evident, for instance Schleswig-Holstein, which in 2015 aimed at 3–400% renewables while North Rhine-Westphalia aimed at 15% (Ohlhorst 2015).

Across the three countries, there are federal-state policy tensions as the central state seeks to shape regional policymaking and regional authorities seek to pursue their own priorities. In each of the case studies, energy and climate policy is a conjoint legislative responsibility. Participation in international forums and agreement-making has tended to establish the national state as the privileged agent in the definition of climate policy, but the active and sometimes divergent role of sub-national states in producing decarbonisation outcomes remains centrally important. Where the national state's commitment to progressive climate policy has waned, sub-national states can be critical to maintaining the momentum of policy action and development, although the level of engagement has varied (Jordaan et al. 2019). In India, for example, the National Action Plan on Climate Change presumes that the sub-national states will contribute to national climate policies (Jörgensen et al. 2015). In Germany, the *Länder* have been particularly crucial in pursuing decarbonisation (Gailing et al. 2013; Moss et al. 2015). Sub-national states in Australia have defined the pace of the engagement with renewable energy, often ahead of the fossil fuel priorities of the federal government (Rosewarne 2022).

In each of the three country studies, regional authorities exercise energy and land planning powers to bolster regional legitimacy. Their energy narratives define new energy regions in the language of co-benefits, from jobs and cheaper and more secure electricity supply to reduced local environmental impact. These are articulated as part of a wider development narrative linked to renewables investment and are integrated into regional development strategies as modes of regulation designed for the emergent renewables regimes (see Moezzi et al. 2017). National governments likewise are increasingly likely to articulate the benefits of energy transition across the broad sweep of development projections, not simply as a side benefit of emissions reduction. As reflected in the 2022 IPCC Assessment Report discussed in Chapter 1, broad-scale economic policy is now closely enmeshed with decarbonisation. This growing move to define decarbonisation as 'opportunity-sharing' (Helgenberger and Jänicke 2017) is directly reflected in regional development planning. The critical issue is how these are correlated to mobilise constituencies and enable an unfolding process of transformation, rather than a one-off adjustment.

Overall, the regulatory packaging of regional renewables enables social legitimacy for the energy transition. In the chapters that follow, we identify and analyse social forces that constitute regional renewables, showing how constituencies and interests are articulated with policy frames, both in alignment and in conflict.

Resulting forms of legitimation vary widely, reflecting divergent regional histories. Energy transition sees regional elites constructing new alliances and alignments for policy, from installation level to inter-regional, federal agencies to international corporate and financial connections. These reflect the pre-existing nexus of social forces, but they also bring something new, as the advent of renewable energy enables the state and others to engage new social forces and strike new 'historic' compromises. Social relations 'on the ground' can be substantially altered in terms of the role of state agencies as well as for landholders and householders. The particular logic of regional development may also be substantially altered, signalling a wider shift in the mode of regulation, especially in terms of socio-ecological relations. In India, Germany, and Australia, these shifts have been decades in the making but, as noted, are far from predetermined. Certainly, in the current period, of an interregnum between fossil fuels and renewables, there is much that is in flux.

Regional Renewable Energy Narratives

Comparative analysis focused on the nexus between renewable energy growth and legitimation can advance understanding of the emerging dynamics of transition. The circumstances for transition, and the barriers to transformation, vary substantially across countries. Countries themselves are composed of widely varying regions, some that have moved to renewable energy, others that remain substantially dependent on fossil fuels. Our research focuses on the former, on regions that have very high levels of renewable energy in the region-level energy mix. In comparing these regions, we aim to gain a closer understanding of the barriers and possibilities for different kinds of energy transition and to highlight lessons for other regions and national states. The structures that enable the prioritisation of renewable energy development, with the identification, negotiation, and approval of installations, along with anticipated linkages in terms of sourcing and energy use, are central to the emergence of the sector and its impacts. The next three sub-sections seek to summarise national-level debates on renewable energy in the three countries, providing context for the ensuing ethnographic chapters.

Karnataka

Since the 2015 Paris Agreement, the Indian government has become an increasingly enthusiastic advocate for renewable energy. At the Glasgow climate conference, Prime Minister Modi announced a target of 500 GW of non-fossil fuel energy by 2030, representing half of India's installed capacity. Ambitious targets, even when they are not fully met, are encouraging the development of large-scale, centralised renewable energy with wide socio-ecological implications. Renewable energy capacity, not including large hydropower, increased sevenfold between 2010 and 2023, reaching 126 GW of installed capacity by May 2023, representing 30% of India's total (Gulia et al. 2022; IEA 2022; Central Electricity Authority 2023).[1]

The rapid expansion of renewable energy in India is shaped by its history of electrification and industrialisation. India's first Prime Minister, Jawaharlal Nehru, famously stated that 'electricity is perhaps the most necessary and the most revolutionary thing which you can take into the rural areas. The moment you take electricity, all kinds of things begin to move' (Kale 2014: 1). Political visions of renewable power articulated by recent Indian central governments have linked with this tradition. Electricity is regulated at both national and sub-national levels, and as noted across India, there is considerable variation in the energy mix between states. As Kale demonstrates, the variations reflect the different histories of political mobilisation around energy.

Renewable energy targets are being pursued through a national policy framework that is focused on developing renewable energy in key states with high renewable energy potential. The overall 500 GW target will largely be met by 10 of India's 28 states (Doczi 2021). The central government seeks to meet its renewable energy targets through its Renewable Purchase Obligation (RPO) policy. RPOs primarily apply to state-level electricity distributors (known as DISCOMS) and require them to source a certain share of their electricity from renewable energy generation (Ministry of New and Renewable Energy 2023a). State-level DISCOMS, many of which are publicly owned, account for 80% of wholesale demand for electricity in India, with corporate PPAs accounting for the majority of the remainder (Gulia et al. 2022: 10). The national government has also channelled investment towards expanding and upgrading the national transmission system to enable access to renewable energy and also to allow inter-state trading of electricity produced by renewables. These policies culminated in the announcement of a national 'Green Energy Corridor' in 2020 across eight central India renewable-rich states, funded by the central government (Ministry of New and Renewable Energy 2023b).

While wind power dominated the early development of modern renewable energy in India, the more recent rise of solar power has been particularly dramatic. In 2010, India had almost no solar power. By May 2023, it had installed

[1] Installed capacity differs from electricity generated. Installed capacity is the maximum level of electricity that could, technically, be produced from a given installation. Electricity generated is the amount of electricity that is actually produced over a given period of time. Renewable energy technologies tend to have lower 'capacity factors' than fossil fuel technologies, meaning they run at maximum capacity less often. Thus, the proportion of installed capacity of renewable energy is often higher than the proportion of electricity generated by renewable energy.

67 GW (IEA 2022; Central Electricity Authority 2023). This rapid increase was facilitated by the Jawaharlal Nehru National Solar Mission, which was launched in 2010 to promote solar energy as a major foundation of energy security, climate action, and access to electricity. As noted in the original Solar Mission policy, the decentralised possibilities of solar would 'ensure that people with no access, currently, to light and power, move directly to solar, leap-frogging the fossil fuel trajectory of growth' (SECI 2010: 4). Prime Minister Narendra Modi leant heavily on the promise of solar in his successful 2014 national election campaign as a basis for expanding electrification to rural areas, pledging that every Indian household would have access to at least one indoor light powered by solar by 2019 (Katakey and Chakraborty 2014). While electrification of Indian households has increased rapidly, this promise was not met (Agrawal et al. 2020). Nonetheless, targets for solar continued to feature heavily in Modi's campaigning, such as in his 2024 re-election campaign where he announced a 'PM's Sun Home' (*PM-Surya Ghar*) scheme of rooftop solar subsidies promising 'free electricity' for 10 million households (Ministry of New and Renewable Energy 2024a).

The Solar Energy Corporation of India (SECI) is the central government-owned enterprise tasked with implementing the National Solar Mission. It has sought to achieve targets by promoting large-scale solar parks, working with state governments to secure access to land, facilitating grid connections, and financially guaranteeing projects. Though SECI also has off-grid and rooftop programmes, it views large-scale solar parks as a way of concentrating transmission infrastructure and overcoming land bottlenecks that have emerged in more distributed generation models. Land issues are especially acute in India due to the close connection between land and livelihood, leading to potential disputes where there are competing claims to access and ownership, and where renewable energy is perceived to compete with productive, or common, agricultural land.

To attract renewable energy developers to invest in large-scale solar parks, SECI has used competitive bidding mechanisms to auction PPAs that lock in future sales of electricity. SECI has awarded PPAs to 47 solar parks with a combined capacity of 25 GW via its central auctions at rapidly declining prices (IBEF 2021). Bids can be for identified solar parks or as open bids for developer-driven generation opportunities. SECI plays an intermediary role in facilitating PPAs between DISCOMS and generators. As part of the contracts with generators, it offers a guarantee, backed by the Reserve Bank of India, against non-payment by financially distressed DISCOMS (Kenning 2017). SECI necessarily collaborates with state authorities because the Indian constitution lists electricity as a 'concurrent responsibility' between both levels of government and land (other than forest land) as a state responsibility.

The acceleration of renewable energy development in India is occurring in the context of an overall expansion on electricity demand-driven by processes of electrification and industrial development. In this context, SECI's aim is to promote the expansion of solar while not undermining existing coal, which it views as providing an important source of 'flexible' generation. New coal is expensive and international private finance and state and multilateral development banks are facing pressures to reduce their investment in coal. Most proposals for new coal-fired power plants are never commissioned. However, increasing demand also means that dynamics seen in Australia and Germany where renewable energy development results in early retirement of coal assets are less present in India.

The central government facilitates investment in renewable energy through its 'blanket' environmental approval of all solar and wind energy projects. This blanket approval exempts renewable energy projects from undertaking Environmental Impact Assessments and gaining consent from relevant State Pollution Control Boards (although forest and national security approvals are needed) (Kumar and Thapar 2017). Blanket environmental approvals are designed to fast-track renewable energy development but, as we will document, remove an important lever for communities to secure better socio-ecological outcomes from solar and wind. Indeed, research documenting environmental pressures and social dislocation from large-scale wind and solar projects across India is proliferating (Yenneti et al. 2016; Lakhanpal 2019; Mookerjea 2019; Bedi 2022; Singh 2022; Shokrgozar and Girard 2024). Much of the pressure on land arises from heightened rivalry between sub-national state governments to attract renewable energy capital: as Yannati et al. find, as 'the competitiveness between India's sub-national states intensifies to capture foreign and national investors for solar developments, more and more common property resources and farmlands are being acquired for that purpose through "fast track procedures"' (Yenneti et al. 2016: 98).

National-level policies have combined to attract private investment in India's solar sector from both Indian and foreign private energy companies. India's renewable energy generation is dominated by private ownership, in stark contrast to the coal-fired power sector. Sixty-five per cent of coal-fired power generation is owned by either state or central governments, with 35% private ownership. In contrast, 97% of renewable energy (excluding legacy hydro) is privately owned, with 3% in public hands (Central Electricity Authority 2023). Ten companies own over half (56%) of installed renewable energy capacity. The biggest owner is Adani, followed by Azure, NTPC Limited, ReNew, Tata, SB Energy, Acme, Avaada, SJVN, and Ayana (Gulia et al. 2022: 12). Investment from international capital is commonly organised through a joint venture model with Indian-based developers. Assets are then regularly on-sold through a BOOT model of 'build own operate transfer'.

Karnataka's Energy Transition

The region that is now the state of Karnataka was one of the earliest in India to electrify. The government of the colonial era Princely State of Mysore, which largely overlaps with the territory of modern Karnataka, developed a system of state-owned hydropower in the early twentieth century as a strategy for developing its industrial base and raising state revenue (Kale 2014: 37). This system of state ownership was carried on following independence through the framework for state-owned enterprises established by the Electricity Supply Act of 1948 (Kale 2014: 40–42). In addition to supporting industrial development, electrification was a state response to a strong rural movement demanding electricity for irrigation. The Karnataka government used a model of cross-subsidisation between urban and rural energy users to electrify all irrigation pumps by the early 1990s, making it one of the first two states to achieve this, alongside Maharashtra (Kale 2014: 90). Following successive market reforms in the 1990s and 2000s, the Karnataka government embraced the potential for private generation – first coal, then renewables – to support the growth of its IT hub city of Bengaluru (Kale 2014: 127).

Karnataka is now one of the leading renewable energy states in India. Fifty-four per cent of installed capacity in May 2023 was renewable energy (not including large hydropower). This is compared with 31% coal, 12% hydropower, 3% nuclear, and no gas-fired capacity (Central Electricity Authority 2023). In total, Karnataka had installed 16.7 GW of renewable energy capacity by 2023. This places Karnataka as the Indian state with the fourth highest installed renewable energy capacity, behind Rajasthan, Gujarat, and Tamil Nadu, which make up the top three, respectively.

Karnataka's initial renewable energy development trajectory was dominated by wind energy, but wind has since been overtaken by the rapid installation of solar. Wind power was steadily installed since the mid-2000s and had reached 5.3 GW of installed capacity in May 2023. Since 2016, when there was only 100MW of solar power installed in the state, Karnataka has emerged as a leading state for solar power in India, reaching 7.6 GW in May 2023 (KREDL 2023c). Karnataka's solar power is overwhelmingly large-scale (92%), with only 600 MW of rooftop solar and 30 MW of off-grid (Ministry of New and Renewable Energy 2023b). The rapid growth of solar can be attributed to much larger-scale developments, such as a Pavagada solar park, supported by land aggregation by long-term PPAs, that we discuss in Chapter 3. The policy innovations overcame some of the bottlenecks that were experienced by the wind industry, which is evident in the high level of cancelled wind projects compared with solar (KREDL 2023d).

The growth of renewable energy in Karnataka has been driven by a pro-renewables policy framework implemented by the state government. The state has become one of the most active sellers of surplus power on the Indian Energy Exchange platform for inter-state trading (Deepika 2022). The policy framework has meant that Karnataka is the only state in India that has exceeded its national RPOs. It has consistently exceeded state-level targets, such as its target of 6 GW of solar energy by 2021, which it easily met. Its most recent renewable energy policy for 2022–2027 aims for an addition 10 GW of renewable energy installation (Government of Karnataka 2022).

The state level has been a scale of policy experimentation in India's energy transition, and Karnataka has played a leading role (Elavarasan et al. 2020). In other states and territories, examples include mass access to feed-in tariffs for farmers using solar water pumps in Gujarat, aggregated rooftop solar in Delhi, extensive use of small-scale wind power in Tamil Nadu, numerous examples of community-level power for micro-grids (Bhattacharyya 2019), and, as we outline in Chapter 3, the use of land leasing arrangements and competitively tendered PPAs for large-scale solar power in Karnataka.

The Karnataka state government outlined its vision for renewable energy development in a series of 'Karnataka Renewable Energy Policy' documents, the latest covering the period of 2022–2027 (Government of Karnataka 2022). These documents contain headline goals including an aim to become 'the leading renewable energy State in the Country', an 'investor-friendly state' for generation and transmission, developing export capacity to other states, the 'solarisation' of agriculture, the creation of an energy storage market to stabilise the grid, and to spread renewable energy capacity across the different geographical regions of the state. The vision is for renewable energy to mimic the perceived benefits of coal in terms of capacity utilisation, reliability, and dispatchability, but on a lower cost basis. Through these documents, four key reasons emerge as to why Karnataka has enthusiastically pursued the development of renewable energy.

First, the state has favourable solar and wind conditions that create a large renewable energy potential. It is estimated that Karnataka's installed capacity is only 11% of the total renewable energy potential in the state (Michael et al. 2023: 11). In contrast, Karnataka has no coal mining industry. As a result, it is dependent on coal imports from other states, including Maharashtra, Odisha, Telangana, and Jharkhand, and has faced coal shortages due to its exposure to increasing transport and spot prices (Buckley and Shah 2018: 8).

Second, the state had beens faced with a growing power supply deficit, with shortages at peak times, making the state a net importer of energy. The problem was getting worse through the 2000s and 2010s as demand for energy grew and renewable energy was seen as offering a means of alleviating the problem.

The dynamics of growing demand for energy are a distinct feature of Karnataka compared with the Australian and German cases. Across all of our contexts, electrification is projected to drive increasing demand for electricity. However, in an emerging economy context such as Karnataka, electrification satisfies a still unrealised demand for energy, rather than simply a switch from fossil fuel-based non-electric energy to renewable energy-powered electricity. Nonetheless, by 2022, due to the addition of new renewable energy capacity, Karnataka had effectively become self-sufficient, producing 99.9% of its total power supply needs (Michael et al. 2023: 22).

Third, the state aimed to attract private investment from private capital. Clean energy has been viewed by successive governments as important for state growth, both as a direct source of investment and for the supportive role played by electricity generation in other growing areas of the economy, such as the IT industry. As with the rest of India, Karnataka has relied on private investment for developing renewables. Ninety-nine per cent of renewable energy capacity in Karnataka is privately owned. In comparison, 78% of coal-fired power generation is owned by state and central governments (Central Electricity Authority 2023).

Fourth, there was a need to reduce costs of supply to address financial stresses faced by embattled state-owned electricity supply companies (DISCOMs). As part of market reform processes, like in other contexts such as Australia, Karnataka unbundled generation and distribution. Distribution was split between five DISCOMs created in 2002–2005 out of former Karnataka Power Transmission Corporation Limited (KPTCL), with the Bangalore Electricity Supply Company Limited (BESCOM) being the largest. The Gurucharan report into Karnataka's electricity distributers released in 2022 summarised how this context was resulting in financial crisis for DISCOMs, stating that the DISCOMs 'as a whole, are loss-making and debt-ridden' and had become reliant on regular state bailouts (Gurucharan 2022: 2). Deficits are the result of per unit costs exceeding per unit revenue, creating a situation in which 'the more they sell the more they lose' (Gurucharan 2022: 12). DISCOMs have, in turn, been financially constrained by credit ratings that have been downgraded due to missed payments to generators and lenders.

The reasons for these financial difficulties have been heavily contested but are framed in terms of the failure of distribution companies to adhere to the accumulation logics of now corporatised government services. Various actors and analysts blame a combination of the subsidised price structure, the prevalence of unmetered consumption, and electricity theft. According to the Gurucharan report, 45% of energy consumption is unbilled (Gurucharan 2022: 21). This relates not only to the lack of revenue from metered power but also to 'pilfering' (or illegal electric connections), which is not policed or punished due to the ongoing political power of constituencies that believe in the right to free electricity as a public good. Farmers receive largely unmetered power for irrigation pumps, generating no revenue. Around 30% of power input into the grid is provided for free for agricultural purposes, as well as rural electrification programmes in villages, reflecting development goals but also the historical power of farming constituencies (Gurucharan 2022: 16). While the progressive pricing structure means that poorer households and farmers are most subsidised, 75% of electricity sold gains some kind of subsidy, meaning that the overall price paid is below the cost of supply (Gurucharan 2022: 11). Other industrial and commercial users facing higher tariffs are in turn responding by striking private contracts for captive power with independent power producers (including renewables), which reduces the higher-paying customer base for the distributors. Adding to price subsidies and unmetered electricity, distribution companies preside over an ageing infrastructure subject to extensive leakage.

The expansion of renewable energy offers a partial solution to many of the problems facing DISCOMs, reducing the need for other structural reforms to address these issues. Moving to renewable energy substantially decreases the cost of daytime energy and is seen as helping to reduce the deficit. The increase in renewable energy has significantly reduced power costs for distribution companies. In 2020–2021, the average cost of electricity from thermal coal power stations was between 6.96 Rs./KWh for Karnataka's state-owned generators, 4.57 Rs./KWh for central government-owned generators, and 8.4–9 Rs./KWh for Adani's Udupi power station. This compared with an average of 4.41 Rs./KWh for renewable energy sources (Gurucharan 2022: 18). As we outline in Chapter 3, the newer solar installations, such as Pavagada, are providing even lower cost power, at less than half the average rates of older renewable energy generation.

Karnataka's renewable energy goals have been pursued through a policy framework that is overwhelming geared towards attracting private investment by making renewable energy an investable and bankable asset (Government of Karnataka 2022). Karnataka has implemented various policy frameworks that have been focused on two main areas: securing the land needed for the installation of renewable energy projects and the market conditions needed for renewable energy developers to sell electricity. The state government has developed regulatory frameworks for land procurement via land acquisition or leasing. This has included changes to enable the automatic conversion of agricultural land for industrial purposes, with no other planning permission required other than from the Department of Energy. With this conversion, renewable energy development is entitled to a range of industrial policy subsidies and tax concessions otherwise targeted towards manufacturing,

The state government has developed policy mechanisms to underwrite revenue for generators selling power from renewable energy projects. Chief among these instruments are government-organised reverse auctions, which competitively tender PPAs on behalf of DISCOMs. These mechanisms have been particularly important for solar power and have led to a high percentage of planned installations being actually commissioned, compared with the use of standardised feed-in tariffs for older wind installations. Up to May 2023, KREDL, the Karnataka Government's renewable energy agency, had implemented eight phases of competitive bidding for solar PPA, supporting 2,279 MW of installed capacity across 149 installations. This was in addition to the separate competitive tendering process that was organised to auction PPAs for the eight allotments in the 2,050 MW large-scale Pavagada solar power, which we discuss in more detail in Chapter 3 (KREDL 2023b). The state government of Karnataka has also supported what it calls 'open access' for power producers and consumers to bypass DISCOMs and negotiate power supply contracts directly. This is often a 'captive' arrangement where independent power producers supply large industrial energy users directly or where private parks feed into the grid. By May 2023, 2,227 MW of renewable energy across 104 projects had been commissioned in this way (KREDL 2023b).

Karnataka's pro-investor renewable energy policy framework has attracted various types of investors into the market, including Indian conglomerates, independent power producers (both Indian and foreign), institutional investors (e.g. pension funds and infrastructure funds), private equity, and foreign state-owned companies. The solar sector has been more internationalised than the wind sector. In Karnataka, private ownership of renewable energy generation is far more concentrated in the solar industry than the wind industry.

Ownership of renewable energy assets is constantly in flux, as new developments proceed and existing assets move across balance sheets through mergers and acquisitions. As of 2021, the top 10 biggest owners of solar together owned 58% of total capacity. For wind, the equivalent figure was 30%. As of 2021, the biggest owner of both solar and wind capacity was ReNew, which owned 12% of solar capacity and 6% of wind capacity. ReNew developed as an independent power producer focused on renewable energy, which initially attracted investment from Goldman Sachs. In 2021, it merged with a Special Purpose Acquisition Company to form a UK-based parent company and was publicly listed on the NASDAQ, and in 2022, the Canadian Pension Plan Investment Board became its biggest shareholder (Barman and Pathak 2022).

Overall, large-scale capital has been far more attracted to Karnataka's solar industry than wind industry. The other largest investors in Karnataka have included Indian conglomerates Adani, Tata, and Hero, and internationally owned companies Fortum (Finland) and Softbank (Japan). Wind owners are more likely to be small companies; however, Indian conglomerate Atria and Hong Kong-owned CLP (owner of Energy Australia) have invested in Karnataka's wind sector.

Through ownership stakes in these energy and industrial conglomerate owners, international investors have exposure to Karnataka's energy sector. For example, pension funds including the Quebec Public Pension Fund have ownership stakes in Karnataka renewables via their significant shareholdings in Azure (invested in solar) and CLP India (invested in wind) (KREDL 2023a).

Within Karnataka, renewable energy development has been geographically uneven but more distributed than coal power. Karnataka is comprised of 30 districts. All thermal generation is located in four of these districts: Vijayapura, Ballari, Raichur, and Udupi. In contrast, all 30 districts have some renewable energy capacity commissioned (including small hydropower, biomass, and co-generation). Twenty-six out of 30 districts have wind, solar, or both installed. However, 47% of renewable energy capacity is concentrated in four of these districts: Tumkur, Belagavi, Vijayapura, and Chitradurga. One of these districts – Vijayapura – is the site of coal-powered generation, while two of the other coal districts – Ballari and Raichur (the most coal-heavy district) – also host 5 and 4% of Karnataka's renewable energy, respectively (Buckley and Shah 2018; KREDL 2023e).

In general terms, India's 'energy transformation' has a scope and reach that far exceeds the parallel developments in Germany and Australia. The sheer scale is unfathomable as a country of 1,400 million people, as against Germany's 83 million and Australia's 26 million. Karnataka's population alone, at 64 million, is more than double that of Australia's. The scale of population is matched by the sheer scale of energy needs and the size of energy installations. No wonder that one of the world's largest solar farms was built in Karnataka and that its successor will now outsize any other solar farm by a factor of two. The fact that India is an industrialising country with a rapidly growing energy demand is also a key point of difference. Despite this the themes are remarkably familiar, notably in terms of issues of climate policy, neoliberalisation, and regulation. As outlined in Chapter 3, these themes are played out in the substantively very different context of port-colonial developmentalism, with differing results yet with parallel dynamics.

Brandenburg

The basis of German energy transition policy can be broadly divided into three phases: an initial phase in the 1990s following the introduction of the first feed-in tariff in 1991, a second phase from 2000, when the Renewable Energy Act (EEG) was introduced, and a third phase from 2011, when the Fukushima disaster in Japan marked the second nuclear phase-out and the *Energiewende*, the energy transition, became the 'official headline of the new German energy paradigm' (Fabra et al. 2015: 5). Across this period, wind power takes the lead; solar power grows at the same pace from 2010, from a much lower starting point (see Figure 2.1).

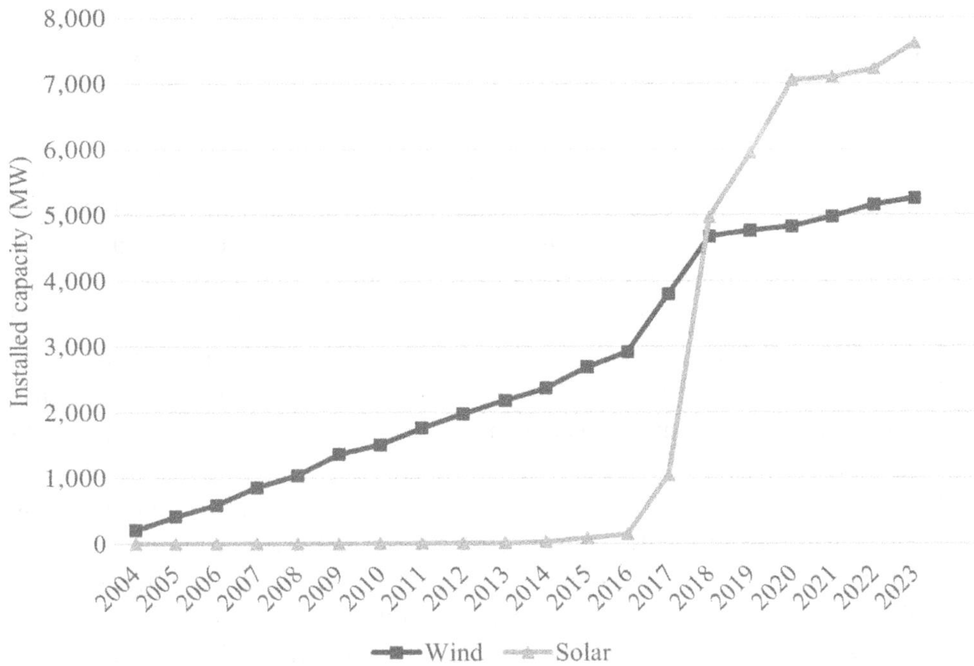

Figure 2.1 Installed wind and solar power in Brandenburg 2004–2023 (MW) (Source*:* Brandenburg State Government)

The German energy transition can be seen as a social process that originally emerged from below, having roots in the anti-nuclear movement of the (then) Federal Republic of Germany and the oil crisis of the 1970s, and what Byzio et al. (2008: 5) describe as a 'utopian vision of "soft" energies and decentralised economic and social structures'. Out of this vision arose a gradual process of technical innovation, which provided the basis for the 'take-off' of renewable energy in the early twenty-first century. Prior to this take-off, the newly unified Federal Republic of Germany introduced the first Electricity Feed-In Act, a law that attracted little attention when it came into effect in January 1991 (Wettengel 2020). The Act ensured for the first time that utilities needed to absorb into the grid electricity generated by hydropower, wind energy, solar energy, landfill gas, sewage gas, and biomass and set a price for remuneration at 75–90% of the average end consumer price per kWh. Together with other programmes, such as the 100/250 MW programme, providing grants for installation and operation of wind turbines, or the 100,000 roof programme with subsidies for the installation of roof-top solar panels, it provided a base for the development of renewable energy and enabled the rapid penetration of wind energy into the German energy market by the turn of the century (Sijm 2002).

The second phase of German climate and energy policy is marked by the coalition agreement between the Social Democrats and the Green Party, which enshrined 'sustainability' as the core principle of the coalition's 'Agenda 21', with a goal of 'economically competitive, socially just and ecologically sustainable development' (Lexikon 2015). It also explicitly referred to ecological modernisation as 'the great chance to protect the natural foundations of life and create more jobs' (Lexikon 2015: 1). Ecological modernisation conceptualises innovation-driven economic growth as a means to reduce the environmental impact of production and consumption, thereby transforming the trade-off between ecology and economy into a positive-sum game (Hillebrand 2013: 665). Ecological modernisation can be seen as a core principle of Germany's energy policy narrative over the last two decades, underpinning the creation of a 'booming environmental industry' (Hillebrand 2013: 668), which developed through federal and state legislatures regulating the expansion of renewable energy and assisting the process of capitalisation described earlier.

In 2000, the coalition government passed laws initiating a phase-out of nuclear energy. It also passed the Renewable Energy Act (EEG) in 2000, which mandated 'grid priority' for renewables: grid operators were obliged to buy electricity and biogas generated by renewable energy producers at a fixed price and privilege it when feeding energy into the grid (Bundesgesetzblatt 2000). The Act also expanded the size of renewable plants that fell under the EEG and its attractive and calculative remuneration scheme (i.e. the feed-in tariffs), to serve the target of at least doubling the share of renewable energy in the country's overall energy consumption by 2010, that is, to roughly 6% of the primary energy consumption. The Act also intended to put into practice Germany's commitment to reduce greenhouse gas emissions by 21% by 2010 under the Kyoto Protocol. The EEG and its alternations and expansions in 2004 and 2009 contributed to a substantial increase in installed capacity of renewable energy, increasing from 4 GW in 1990 to almost 80 GW in 2012 (Umweltbundesamt 2023b). It fostered the installation of wind offshore and onshore, biomass plants, and solar as well as a significant solar industry for research, development, and production, leading to Germany developing an image of *Vorreiter* or 'trailblazer' in climate and energy policy (BMWK 2014). Its various mechanisms and details, however, also contributed to an increase in the electricity price for end consumers.

The third phase of the German energy policy was ushered in with laws introduced after the coalition of Christian Democrats and Liberals (2009–2013) had revised and prolonged the nuclear phase-out. The Fukushima nuclear accident was the turning point for the 'phase-out of the phase-out' and the catalyst for the nuclear law of 2011, that now scheduled the complete exit of nuclear power in Germany for 2022 (Bundesgesetzblatt 2011; Spiegel 2011). At the same time, the German government made the Energiewende a matter of national political concern, setting new targets of 35% of national electricity generation to be generated from renewable sources by 2020 and 80% by 2050 (Bundesregierung 2011). However, the targets along with the nuclear law did not lead to an increase in renewable energy production, but initially to an increase in brown coal production, from 176 million tons in 2011 to 183 million tons in 2013 (Statista 2024).

Consequently, greenhouse gas emissions also rose from 919 million tons in 2011 to 941 million tons in 2013. They started to decline again only in 2014 (Umweltbundesamt 2023a). In 2014, the government also passed the next amendment of the EEG, following a longer reform discussion in 2013. The EEG 2014 significantly altered rules and regulations for renewable energy production, including caps and reduced remuneration for biogas; taxing self-consumption of solar electricity above 10 kW; competitive bidding for large-scale solar; and reduced feed-in tariffs for on-shore wind in wind-strong areas (Leiren and Reimer 2018). It also scheduled the path towards reverse auctions, which were made mandatory with the EEG 2017. The aim was to shift away from incentives to contracts. Effectively, the EEG 2014 and 2017, criticised by many stakeholders as jeopardising expansion in general, and citizen participation in particular, led to a decline in wind installation after the EEG 2014, and to a plunge in 2018 (BEE 2013; Windbranche 2023a). There was a stabilisation of wind installation at a low level in 2021 and a slow increase in 2022 (Windbranche 2023a).

In 2020, the previous German federal parliament passed the ‘coal exit law’ (Kohleausstiegsgesetz), legislating a phased decommissioning of coal-fired power plants and the final shutdown of all coal-fired electricity generation in Germany by 2038. In October 2021, the coalition agreement signed by the incoming Federal government of Greens, Social Democrats, and Liberals (Free Democrats) set the achievement of the Paris climate protection goals as a top priority (Koalitionsvertrag 2021–2025: 5). To this end, the coalition agreement outlined a reliable and cost-efficient path to climate neutrality by 2045, with 65% CO_2 savings by 2030; it also stated that planning and approval procedures for renewable energies should be streamlined so that the expansion of wind and solar energy can be accelerated (Koalitionsvertrag 2021–2025: 44, 70). The share of renewable energies is to rise to 80% of electricity consumption by 2030; photovoltaics is expected to reach a capacity of 200,000 MW countrywide in 2030, including a solar obligation for commercial new buildings; for onshore wind energy, the coalition agreement specifies a requirement for 2% of the country’s surface area to be set aside for electricity generation from wind. A related federal law was passed in 2022 with the aim of improving the availability of land and simplifying the planning procedures for the accelerated expansion of wind turbines by setting binding land use targets for the federal states – of 1.7% by 2027 and of 2% by 2032.

This new push to accelerate the German energy transition is, in no small part, a reaction to its recent slowdown. Germany almost failed its self-imposed greenhouse gas emission reduction goal of 40% by 2020, and only barely succeeded in reaching it due to the Covid-19 pandemic and the related decline in individual transport and transport of goods.[2] With a share of 45.4% of electricity generation from renewable energy sources, the country hit a new record in 2020 (though this remained static into 2022) (Umweltbundesamt 2022: 6). The electricity generated from renewable energies even slumped by about 7% compared to the previous year and has thus fallen from 251.1 billion kWh in 2020 to 233.6 billion kWh, due to not particularly windy months, especially in the beginning of the year 2021 (Umweltbundesamt 2022: 7).

Similarly, the amount of newly installed onshore wind capacity in Germany has been falling sharply since the amendment of the Renewable Energy Law in 2017. It had been increasing steadily since the early 2000s, reaching a peak in 2017, and rebounding only from 2020 onwards, but nowhere near enough to reach 2017 installation levels (Deutsche Windguard 2023). The expert commission 'Energy of the Future', which monitors the progress of the *Energiewende*, sees 'very considerable challenges' for further expansion of onshore wind, naming lengthy bureaucratic approval processes, the increasing frequency of legal challenges to such approvals, and clarification of problems surrounding species protection and aircraft flight paths as just some of the issues that urgently need to be addressed (Löschel et al. 2021: 25). The abovementioned 2022 wind-energy-on-land law was an attempt by the federal government to tackle some of these issues; however, relevant federal authorities will not assess the success of this law in tackling bureaucratic hurdles and addressing the uneven distribution of wind energy installations across individual states, until 2027. However, as the Expert Commission noted, strengthening the social acceptance of wind energy installations *in the countryside* – and hence in places where renewable energy production takes place – will be decisive for achieving the necessary expansion of wind energy and, with it, Germany's climate and energy goals (Löschel et al. 2021).

The latest amendment to the EEG in 2023 also states that installation and operation of renewable power plants is of preeminent national interest, thereby aiming at faster administrative processes while restricting environmental protection. According to recent analysis by the Federal Office of the Environment (Umweltbundesamt), however, Germany will not meet its current emissions targets without further policy measures (Umweltbundesamt 2023b). The country only met its 2020 emission reduction targets thanks to the Covid-19 pandemic and the resultant decline in industrial production, traffic, and transport (Deutschlandfunk 2021). The Russian war on Ukraine has put further pressure on the German energy system, facilitating new fossil fuel investment and infrastructure lock-ins, including

[2] www.deutschlandfunk.de/erfolgsmeldung-mit-schoenheitsfehler-deutschland-erreicht-100.html.

also fast-tracked installation of LNG terminals, arguably beyond needed capacity (Holz et al. 2023), and the signing of contracts for (fracking) gas imports.

A further infrastructural barrier to accelerating the energy transition (as in Australia and India) is the electricity transmission grid, whose expansion has been identified as an urgent priority for the integration of more renewable energy into the German energy system. Expansion of the extra-high-voltage grid and high-voltage grid is the responsibility of four large transmission network operators (TSOs), TenneT, 50 Hertz, Amprion, and TransnetBW, with 50 Hertz being responsible for the east and northeast of Germany. Distribution network operators take care of the regional and local power grids on medium-voltage and low-voltage levels; Germany currently has 872 of these network operators.

The German Energy System in European Context

At the same time, Germany sits at the heart of an interconnected European electricity system. Because of its size and central geographical situation within Europe, it is an important player on the European electricity market and a hub for Europe-wide power flows, as well as having to meet EU policy settings. Germany is also exporting more and more electricity to neighbouring countries. Import and export flows are driven by the wholesale prices on national electricity exchanges, which are influenced by the respective demand for electricity, the amount of electricity generated from renewable energy, and the fuel costs for conventional power plants. Electricity is physically exchanged with nine direct neighbouring countries – Denmark, the Netherlands, Luxembourg, France, Switzerland, Austria, the Czech Republic, Poland, and Sweden (via a submarine cable).

Yet, as regards total energy use, Germany is dependent on imports from abroad. According to Eurostat, the share of net imports in energy consumption was 69% in 2021 (Statistisches Bundesamt 2024a). Despite the expansion of renewable energy, the supply of gas continues to play an important role for industry and private households in Germany. As a result of the Russian war in Ukraine, the German government has ended its energy dependency on Russia, not only for gas but also for coal and oil. Since September 2022, no Russian natural gas has been delivered directly to Germany via pipelines. Instead, Germany gets increased natural gas deliveries from Norway and the Netherlands as well as additional LPG imports (Bundesregierung 2023).

The dependency on energy imports of the EU as a whole was 63% (Statistisches Bundesamt 2024a). The EU thus covered more than half of its energy requirements from imports. In order to eliminate dependence on Russian imports for future energy generation, the EU Commission is planning investments of 300 billion euros by 2030. The plan for solar energy includes the reconstruction of the European solar industry with the aim of quadrupling installed capacity in the EU by 2030 to a total of 600 GW compared to today. The EU-wide power generation capacity from photovoltaics was 162 GW in 2021. Germany had by far the largest share of this with 59 GW or 37% (Statistisches Bundesamt 2024b).

In sum, the development of renewable energy has advanced over the last decades across Europe, as well as in Germany in particular. Renewables in Germany had a share here of 46% of the fed-in electricity in 2022, while wind power has the highest share in the amount of electricity from renewable energy sources with around 24% (Statistisches Bundesamt 2024c). However, several amendments to the renewable energy laws have slowed down anticipated developments in the renewables sector. The switch to reverse auctions in 2017 brought a decline in new installations of onshore wind from 1,853 devices in 2017 to 754 in 2018 and 325 in 2019 (Deutsche Windguard 2019). Smaller and community-based operators could not compete. The subsidies for solar installations meanwhile were capped to a total installation of 52 GW, which also decelerates the speed of installing renewables. Last but not least, the *Energiewende* is not simply a set of policy instruments but also a social process, as we will lay out in more detail in Chapter 4.

The 'Energy State' of Brandenburg

The shift to reverse auctions in the EEG 2017 also affected the German states' ability to influence renewable energy installation, as some states had more ambitious renewable energy production targets than the federal government (Meya et al. 2017): Schleswig-Holstein, for example, was aiming at a 300% share of renewable energy in net electricity consumption by 2025. Brandenburg in 2015 already produced 14.630 GWh of renewable electricity while consuming 18.936 GWh (EnergyMap 2016). Net renewable consumption of 77% increased to 100% in 2021 and 95% in 2023 (MLUK 2024). This figure, however, does not reflect the fact that Brandenburg has, for a long time, been an energy-exporting state: 50% of electricity produced in the state is exported. Reflecting this, Brandenburg calls itself an 'Energieland' ('energy state'), and rightly so (Hettchen 2020). It has significant deposits of brown coal, which have been extracted and burnt for energy generation for more than 150 years (see Förster 1968; Müller 2017), but coal mining and generation are scheduled for shutdown by 2028 and 2038.

The state has been fostering renewable energy installation for several decades. In the 1990s, the government started the first programmes to financially support renewable energy production. In 2008 and 2012, the state adopted the programmatic 'Energiestrategie 2020' and 'Energiestrategie 2030', respectively. The Energy Strategy 2020 scheduled an expansion of installed wind capacity to 7,500 MW till 2020 and 10,500 MW till 2030. Considering that most of the growth from 2020 on will be implemented through repowering of older wind power plants, the state also inserted in the later strategy an aim of about 2% of the land (585 km²) designated to wind energy production (Overwien & Groenewald 2015: 603). By early 2023, Brandenburg had reached a total of 489.2 MW installed biomass capacity, 6,600 MW solar electricity capacity, and 8.660 MW installed wind capacity, equalling 575 biogas plants, 103,183 PV systems, and 4,057 wind turbines (Bundesnetzagentur 2024a).

The Energiestrategie 2030 was revised and updated in the Energiestrategie 2040 and passed in 2022. Its aims include to be climate neutral by 2045; to reduce primary energy consumption through efficiency by 39% by 2040; to increase the share of renewables in primary energy consumption to 68–85% by 2040; to reach 100% net renewable electricity consumption by 2030; and to install at least 15GW wind energy and 33 GW PV by 2040 (MWAE 2022). Unlike the Energiestrategie in 2020, the 2030 and the 2040 strategies also explicitly include forms of social and community participation. The state describes its strategy as an 'energy-political target square', comprising environmental and climate compatibility, profitability, security of supply, and acceptance and participation (MWAE 2022). To reach these goals, Brandenburg set up a package of measures to make the use of energy more efficient, to reduce CO_2 emissions in conventional heat- and electricity generation, to improve transmission, distribution, and storage, and to expand renewable development. Arguably also as a consequence of promised tax benefits not reaching affected communities, as well as grid charges impinging on individual electricity consumption bills, Brandenburg's government passed a law in 2019 that obligates wind energy operators to pay 10,000 Euro per year per wind turbine directly to all adjacent communities within a 3-kilometre radius. The bureaucratic and participatory framework and practice of installing wind and solar energy production will be laid out in detail in Chapter 4.

The policy regime supporting the energy transition in Brandenburg has been oriented towards supporting private investment. As a result, the more established wind energy sector in Brandenburg is controlled by the private sector. Currently, almost 100% of Brandenburg's wind farms with more than four turbines are owned by private companies. Renewable energy developers in Brandenburg are typically privately held Germany-based companies owned by a small number of private individuals. About one-third of the 350 operating wind farms are owned by companies based in Brandenburg, one-tenth by companies based in other East German states, and another third by companies in West Germany, France, and Spain. The remaining 20% of wind farms in Brandenburg are owned by companies whose precise location is hard to trace, due to their corporate structure and the use of subsidiary companies and offices. Owners of wind farms are typically specialist companies concentrated on the renewable energy sector, usually with operations across wind and solar. The larger developers have investments and operations beyond Brandenburg throughout other regions of Germany and other countries in Europe, as well as in Asia, North and South America, and Africa.

Companies develop wind farm projects both to own them as part of a portfolio of generating assets as independent power producers and to sell them as 'turnkey' projects for other buyers. Buyers of wind farms from their original developers are typically institutional investors (e.g. pension funds), alternative asset managers (e.g. private equity funds), and energy utilities. Many developers continue to generate business from wind parks they have sold by providing operations and maintenance services. The 10 biggest wind farm developers often partner with each other in projects. As of 2020, these 10 had together been involved in developing 68% of renewable energy capacity in Brandenburg. Out of the 350 wind parks and almost 4,000 wind turbines, less than 10 parks were co-owned by locals. The model of 'Bürgerwindparks', or citizen renewable wind energy production, where arguably distributional conflicts would be cut to a minimum, is signally absent in Brandenburg. This trend to corporate renewables, and to greater concentration of ownership in the sector, closely parallels developments in India and Australia.

In sum, as the Brandenburg ministry for economics, labour, and energy states, 'wind energy has developed into an important cornerstone of German energy production and national economy' (MWAE 2024). Brandenburg is among the top three German states as regards wind energy, even though wind development slowed down in Brandenburg after 2017, following the national trend, which was then reversed in Brandenburg from 2021 onwards (Windbranche 2023b). The state is currently experiencing a form of a 'solar rush' (Müller and Pampus 2023), where large-scale solar on converted agricultural land is likely to take the lead in what has been a broader variety of solar plants ranging from roof-top solar of up to 10 kW to open-site installations of more than 100 MWp. Currently, solar park concepts of more than 1,000 GW are being planned and implemented. As Brandenburg has large unused former military areas as well as former lignite mining areas, some of these areas, too, are used for solar power. Current developments, where falling prices of devices allow for an economic set-up and running of large-scale PV without subsidies, however, tend to focus more on investment in converting agricultural land, as it does not require costs for cleaning up contaminations.

Future developments hinge on regulatory contexts and also on the growing financialisation of the sector. Initially, the move to privatise electricity generation, from 1998, took place without any regulatory oversight. In 2005 the Federal Network Agency was created, mandated to promote competition in the sector, principally between the four large generators that accounted for the bulk of energy generation and the range of smaller players. An important step was to 'unbundle' generators from retailers, in order to remove competitive advantage for vertically integrated energy corporates (Pohl 2023b). According to BDEW, there are around 913 electricity network operators nationwide (as of January 2019), most of them local and regional networks, which until 'unbundling' were mostly supplied by a single generator (BDEW 2024).

With unbundling, energy retailers were required to purchase electricity from generators, and most now do so mainly via so-called over-the-counter (OTC) trading and on the power exchange stock market. OTC trading accounts for about three-quarters of electricity sales and has spawned related derivatives, futures, hedge, and spot markets, all designed to minimise risks for retailers in highly volatile but long-term electricity supply arrangements (Pohl 2023b). Electricity trading is now centred on the national Electricity Exchange, the EEX, and is intertwined with speculative trading (ibid.). More than 850 suppliers supply electricity onto the EEX, including power companies, regional providers, municipal and community power companies, and energy traders. The capacity to make a return in these financially volatile 'markets', buffeted by notoriously unstable speculative financial instruments, has become critical for the energy transition. The consequences, for instance, in the context of any unexpected 'shock' to the system, such as the Russian invasion of Ukraine, create their own backwash effects, destabilising the whole system. As in Australia, the neoliberalisation of energy and the purported increase in competition appear to have created more uncertainty for investors. No wonder there has been a rush to more tranquil protected pricing through long-term Power Purchasing Agreements. Nor has the system appears to have delivered price cuts – on the contrary, it is associated with ongoing increases in the cost of energy for households, more than doubling from 1998 to 2022 (Bundesnetzagentur 2024b).

Overall, the Brandenburg experience, within the context of German and European Union energy policy, offers many lessons for energy transition. The issues of on-the-ground social legitimacy are brought into direct dialogue with neoliberal policy and corporate interests. The collision forces important narratives of contestation and renewable energy regulation onto the agenda. The outcomes have shaped and reshaped the direction of the 'energy transformation' in unexpected ways. Chapter 4 takes up this narrative in detail, drawing wider comparative implications.

South Australia

Debate about the challenge of climate change and over the need to reduce fossil fuel emissions was effectively marginalised in Australia from the mid-1990s to the early 2020s. The Federal Labor Government under Paul Keating had made some progress from the early 1990s in joining in international climate negotiations, leading to the Kyoto Protocol of 1997. These limited advances were reversed with the election of the conservative Coalition bloc, composed of the Liberal Party and National Party, in 1996. The Coalition would not concede that there was any substance to climate science, which meant that there was no serious questioning of Australia's reliance on fossil fuels for generating electricity. After forming government in 1996, the Coalition opposed any measures that could restrict the mining and burning of fossil fuels, a position that was forcefully articulated by the Australian delegation headed by the former chief economist at the UN Framework Convention on Climate Change held in Kyoto in December 1997. Government disinterest in climate issues was reinforced by limited public engagement with the issue in the 1990s. As was the case in Germany, environmental movements in Australia were mostly focused on campaigns against clear felling of forests and nuclear disarmament. The focus on nuclear disarmament had particular resonance for South Australian politics with mounting opposition to BHP Billiton's planned copper and uranium Olympic Dam mine at Roxby Downs in South Australia (Goodman and Rosewarne 2015).

There was some very limited movement in 2001 with the creation of the conservative Coalition's 'Mandatory Renewable Energy Target' (MRET). The Coalition was an obstinate critic of claims that emissions were driving climate change and its various leaders expressed this in quite colourful language. The introduction of the scheme in 2001 did not signal a change of heart, especially in light of the government's subsequent withdrawal from the Kyoto Protocol (following the US a year later). It did though suggest a new sensitivity on the issue. The renewable target was set very low, at 2% of energy generation, and was to be achieved through a rebate, not via direct regulation. The MRET was minimalist in the extreme, but it did have the unanticipated (and probably undesired, from a national government perspective) side effect of enhancing sub-national state authority on energy and hence the transition. As such it served as something of a quiet watershed in climate politics and energy governance in Australia. The nation's constitution designates energy governance as a state responsibility and consequently the federal government had to rely on sub-national state governments, several of them led by Labor governments anxious to implement climate policies, now given the role of implementing and overseeing the governance of the MRET scheme.

With this, sub-national states began taking the lead in initiating their own schemes to reduce emissions and encourage investment in renewable energy generation. With the conservative Coalition holding national office for most of this period from the mid-1990s and not wavering from its resolute opposition to effective emissions governance, a deepening antagonism was created between the two levels of government. There was a brief and precarious period of Labor government, 2007–2013, when Australia ratified the Kyoto Protocol and the national government legislated for compulsory emissions reduction measures across all large emitters. Much of this 'Clean Energy Futures' package was reversed with the return of the conservative Coalition in national government from 2013, notwithstanding the retention of some progressive policy measures, which helped underspin significant investment in renewables capacity. One important measure that was retained was the MRET, which labour had expanded tenfold, to a 20% renewable energy target by 2020, with separate schemes for households and utilities (Chapman et al. 2016), all hinging on sub-national state-level implementation.

One consequence of the national-level deadlock, combined with the constitutional devolution of energy policy, was that state governments, and primarily state Labor governments, recognised region-level opportunities and took leadership in charting a path to a more emissions-benign future. The South Australian State Government took the lead in defining this emerging and deepening role for sub-national states. Over the 16 years from 2007 to 2023, South Australia embarked on an energy revolution, increasing renewable sources from just 1% to nearly 70% of electricity generated in the state (see Figure 2.2). This was a product of deliberate sub-national policy, with successive state Labor governments

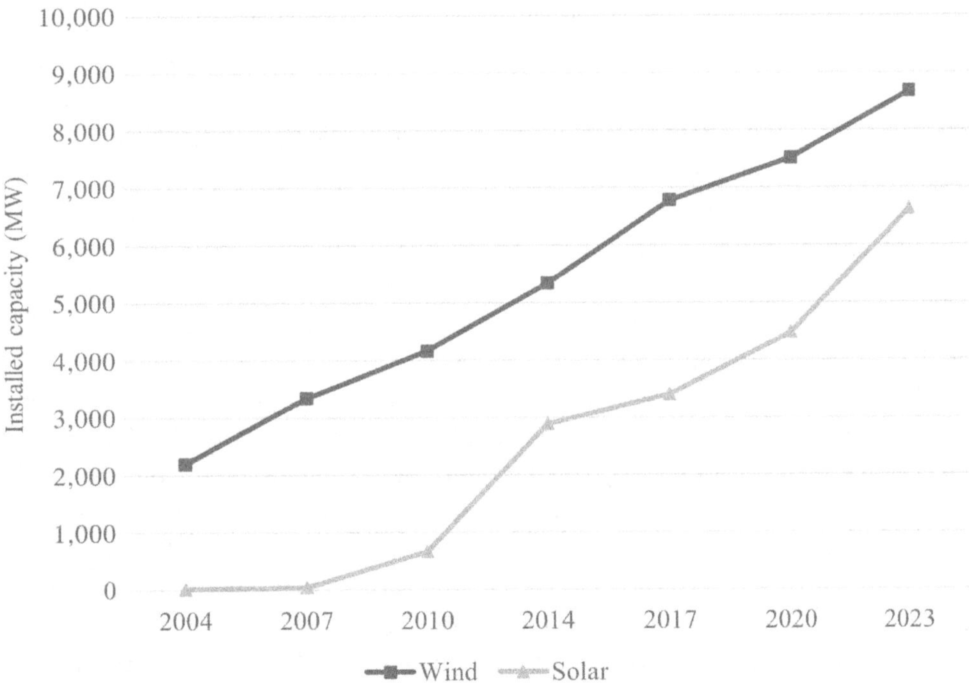

Figure 2.2 Installed wind and solar power in South Australia 2016–2023 (MW) (Source: Government of South Australia)

promoting the development of renewable energy-generating capacity often in defiance of conservative federal Coalition governments. There are two main phases in this remarkable story, first under Rann 2002–2011, and then under Weatherill 2011–2018, and since.

The Rann Government: A Radical Policy Shift

Dependence on fossil fuels began to be questioned in the period following the 1997 Kyoto Protocol, largely led by the opposition Labor Party but also endorsed by the longstanding conservative South Australian government. The election of the Labor Party to government in the state of South Australia in 2002 marked the beginning of a critical shift in advocacy for renewable energy. This was an intriguing development because at the time the main advocates for climate action were environmental organisations, which were, as noted, engaged with several other issues. There was no discernible climate movement pressing the call for action. It would take a second term in office before the Labor government began to think seriously about adopting measures to increase the generation of renewable energy and reduce emissions, and then it was driven by the initiative of one person in particular, Premier Mike Rann.

The scale of the challenge facing Rann was formidable. In 2000, South Australia – like most states in Australia – was heavily reliant on the combustion of coal to generate the state's energy needs. In 2000, the burning of locally mined brown coal at three power plants – the 90 MW Playford A, the 240 MW Playford B, and the 520 Mw capacity Northern Power station, located south of Port Augusta – produced 35% of the power consumed in South Australia. There were also gas units, including the 480 MW Torrens Island A and the 800 MW Torrens Island B installations, accounting for 30% of the electricity consumed in the state. The remaining 30% of the state's power needs were met with imports from the neighbouring state of Victoria (electricity generated mainly from La Trobe Valley coal-fired power stations).

The previous decade had seen the privatisation of South Australia's energy system and the establishment of the National Energy Market (NEM) (Simshauser 2021). The neoliberal ideology that had informed the states' decision to create a NEM and, at least in South Australia, to privatise generation, shaped the parameters for emissions reduction and energy governance (Weller 2018). The Rann-led South Australian government projected its authority on energy in the name of decarbonisation and removed any reticence in questioning the national government and its continued advocacy for fossil fuels as the basis for electricity. This enabled the emergence of a comprehensive, region-level programme to transform the organisation of the energy system. A sharp distinction emerged between South

Australia and other state governments in Australia that were either resolutely opposed to abandoning the reliance on the combustion of fossil fuels or reluctant to commit to strong measures to reduce greenhouse gas emissions.

Rann took his stance against the conservative position on greenhouse gas emissions to the federal level in 2005, calling for a special Council of Australian Governments to discuss climate change (ABC News 2005; Breusch 2005; Rann 2020). From that point on, the South Australian government launched a radical shift in the policy agenda. Rann justified this shift in terms of the pressing need for action on climate and also on the grounds that the Commonwealth Constitution designated states as having principal constitutional responsibility for and legislative authority over energy. In a challenge to the authority of the Coalition government that would have continuing resonance, he asserted the right of sub-national governments to take the lead in confronting the challenge of climate change (The Sydney Morning Herald 2007; Kelton 2010; Rann 2020).

There was substantive progress on the ground. During this period, the state's commitment to transitioning away from fossil fuels catalysed a series of ambitious wind farm projects that would deliver greater generating capacity over the course of multiple development stages. Six wind power farms were constructed in response to the positive support by the state government, the largest at the time being the 91 MW Wattle Point wind farm constructed by New Zealand's Meridian Energy and sold to AGL, the vertically integrated energy company that dominated the South Australian energy market. Its main competitor, Energy Australia, invested in the development of the 131 MW Waterloo wind farm. Australia's third main energy retailer, Origin Energy, sought to lock in energy supplies by negotiating PPAs. One such agreement underwrote the expansion of the Snowtown wind farm, reaching 370 MW; there was further development at Lake Bonney, at 300 MW. Significantly, Rann was not opposed to large-scale emissions-intensive industry, such as BHP Billiton's massive Olympic Dam uranium mine. Instead, he presented these industrial projects as offering opportunities for renewable energy to retrofit and decarbonise through large-scale renewable energy (Parkinson 2011).

Beyond utility-scale installations, Rann also had a vision of sharing the benefits of renewable energy through household rooftop solar. In 2006, South Australia announced its intention to introduce feed-in tariffs for households to encourage take-up. The scheme commenced in 2008 and was declared a success with 8% of power being generated coming from household rooftops by 2015. By 2016, 29% of households were generating energy from solar panels in South Australia, rising to 34%, among the highest rates of take-up in the world (AEMO 2016; Arena 2017); other states followed suit. At the same time, the government sought to lead by example, installing rooftop solar panels on government buildings and committing to meeting 50% of the government's own electricity needs (O'Brien 2009).

The initial focus on the energy sector broadened into an economy-wide agenda. The 2007 report *Tackling Climate Change: South Australia's Greenhouse Strategy*, commissioned and published by the Premier's Climate Change Council, mapped an exhaustive strategy for six sectors: community, industry, energy, buildings, and natural resources, each with goals and strategies to achieve decarbonisation by 2014. The ambitious programme identified the sparsely populated Eyre Peninsula as offering the opportunity to establish a 'green grid' to unlock up to 2,000 MW. At the same time, the government welcomed the closure of South Australia's remaining coal-fired power stations, including the mothballing of the Playford B power plant, and began canvassing other proposals that would effectively block the construction of any new coal-fired plant in the state.[1]

In mid-2009, a sub-department, RenewablesSA, was established to encourage further investment in renewable energy in the state. Rann applauded the proposal by the Labor government led by Julia Gillard to dedicate funds for interconnectors, which would permit South Australia to export green power to other states. Investment in wind farms remained the main focus, with the government using supply agreements to underwrite proposals. Progress was facilitated by funding assistance and access to discounted loans organised by the newly created Clean Energy Finance Corporation and the Australian Renewable Energy Agency (ARENA). In a relatively short time, wind turbines were generating at least a fifth of the state's electricity, from 0% in 2006 to 21% in 2011 (reducing coal's share to 32% from 40% and gas to 47% from 57%) (OpenNEM 2023).

Second Phase: New Impetus under Weatherill Government

A change of leadership in the Labor Party in 2011 refreshed and added new impetus to transitioning, heralding a distinctive second phase in the evolution of energy governance. The new Premier Jay Weatherill, announced that the new government would ramp up South Australia's renewable energy target to 75%. He re-emphasised Rann's insistence on the ability of the sub-national state to play a constructive role in progressing the transition from fossil fuels. Weatherill maintained that states and regions could compensate for national-level failure, when 'national governments find it difficult to reach agreements' (Fedorowytsch 2015). South Australia developed its international profile, joining the States and Regions Alliance of the global Climate Group, which also included Victoria, New South Wales, and the ACT. Nor did the government neglect the local. It sought to focus the commitment to reducing emissions by working with the City of Adelaide to become the world's first carbon-neutral city. This energy policy agenda contrasted with the Coalition's reluctance to actively engage in renewable energy governance, especially if the transition was going to jeopardise the dominance of coal and gas. The extent to which this could translate into open conflict was on full display in a heated exchange when the Coalition Prime Minister and Energy Minister publicly blamed Weatherill's renewable energy policy for causing a state-wide power failure in 2016 following a cyclonic storm that tore down transmission pylons and tripped safety systems (Vorrath 2017).

At the same time, Weatherill sought to create a new wave of investment in the sector through the introduction of large-scale batteries. In 2017, the government established the Renewable Technology Fund dedicating $150 million to support the introduction of battery storage and to explore other storage technologies. Plans to install the world's biggest lithium-ion battery, the 194 MWh Hornsdale Power Reserve to store power from the Hornsdale Wind Farm, gained national notoriety when Elon Musk assured the South Australian public that he could build the battery in 100 days or less and, if he didn't meet that target, he would install it for free (Smyth 2017). Underwritten by the ARENA, the success of the Hornsdale venture spawned wider interest in battery storage to bolster the renewable energy industry and set in train a competition between large-scale battery manufacturers Tesla, Zen Energy, Lyon Power, and Carnegie Clean Energy to supply a share of the expanding market for battery storage. Battery storage, with ever larger capacity, such as Maoneng's Gould Creek battery project, which was approved in late 2021 and has a storage capacity of 450 MWh, was adding considerable firming capacity in renewables supply and offered an effective counter to the Federal Coalition government, which claimed renewables could not provide supply reliability, or 'base load', when the sun was not shining or the wind was not blowing.

Increased storage capacity certainly boosted confidence in wind power. Development approvals under Weatherill brought a second phase of utility-scale wind farms with installations tripling capacity reached under the first tranche. By 2023, there were 23 wind farms in operation and two under construction (Government of South Australia 2023a). Wind farms were generating 44% of South Australia's energy. With battery storage an increasingly conjoint feature enabling the firming of intermittent generation, the contribution that solar farms could make to energy self-reliance was also attracting attention. Established solar farms were responsible for generating 24% of South Australia's energy in 2020–2021, and when the 110 MW Port Augusta Renewable Energy Park and 200 MW Goyder South Hybrid Renewable Power Station are brought into service, solar farms, in conjunction with batteries, will increase the possibility of complete self-reliance and also offer the possibility of exporting power to other states (AEMO 2022; Government of South Australia 2023c).

A significant feature of the rapidity of South Australia's transition from reliance on fossil fuels to wind and solar-generated energy has been the extent of investment by international energy corporations. These corporations have led the increase in the generating capacity of wind and solar farms. Their origins are diverse, from the French energy company, Engie, to New Zealand's Tilt Energy, Germany's Siemens, Thailand's Ratch, Singapore's Nexif, Spain's Ibedrola, and the USA's Lexus. This global interest demonstrated the extent to which South Australia had been able to position itself at the cutting edge of the energy revolution. It also demonstrated the extent to which domestic investment could not be relied on to finance the transition. The input of technologies, from wind turbines to photovoltaic solar panels to batteries, was contingent upon and consolidated the internationalisation of South Australia's transition.

Notwithstanding large-scale corporate involvement, wider public engagement in the transition was also considered an important aspect in the planned development of renewable energy. Again, the focus here was on household solar. With funding from ARENA, a grant of $8 million, and a $30 million loan from the Clean Energy Finance Corporation and, in an agreement with Tesla, the South Australian government established a scheme under which solar panels and batteries would be installed on more than 50,000 homes. The first stage of the scheme focused on the installation at no cost of 5 kW batteries for low-income households living in 1,100 Housing Trust residences (Parkinson 2018b; Government of South Australia 2023a). At the end of 2012, with over 4,000 households with rooftop solar panels connected to the grid and further expansion funded by Tesla, the scheme has become the world's biggest 'virtual power plant'. The government bolstered this with a rebate scheme that would incentivise more households and small businesses to invest in the installation of rooftop panels. By 2023, some 44% of residential demand for electricity was being met by household photovoltaic panels, accounting for almost 20% of the state's total generating capacity: rooftop solar briefly supplied all of the state's electricity needs in 2023 (at 2 pm on 22 September; Parkinson 2023).

One consequence of the uptake in home solar and battery storage systems was the concern among retail energy companies that this trend could result in a substantial loss of their business. AGL, which is the dominant retailer in South Australia, sought to circumvent the loss of custom by contracting with households to sign onto feed-in tariff schemes. This also extended to AGL negotiating the installation of solar panels for households that would then become contracted to feed power into the virtual power plant (Rosewarne 2022). This gave AGL the technical means to draw on household energy to meet supply shortfalls or block flows during periods of excess supply. The Australian Energy Market Operator supported the initiative, viewing it as a 'firming' mechanism, permitting AGL to manage household energy to reduce the potential for network congestion. The arrangement was quite novel and arguably enabled renewables to supply 25% of the evening peak-hour demand (Slezak 2008).

With the imminent closure of the Northern and Playford B power stations in Port Augusta in 2016, a local campaign, Repower Port Augusta, was created to advocate for a just transition for workers into a new renewable sector. A successful lobbying campaign secured funding from the Commonwealth government, a $100 million concessional loan, for the Aurora Solar Project, and in 2017, Weatherill announced that the project would proceed holding out the promise of a new start for Port Augusta; however, as we discuss in Chapter 4, the project had disappointingly failed to secure commercial finance and was dropped (ABC News 2019).

Instability and Continuity

The decision to not proceed with the Aurora Solar Project was a demoralising development and seemed to reflect a change in political sentiment among the voting population as the 2018 South Australian election handed power to the conservative Coalition. The Liberal Party led by Steven Marshall raised the prospect of undoing 16 years of the state-led development of renewable energy. Marshall had indicated his intention to abolish Labor's renewable energy target and cancel the battery contract with Elon Musk (Parkinson 2018). This seemed likely as the federal Coalition government sought to put pressure on Labor states opposed to the exploration of gas deposits and negotiated a $1.08 billion agreement with the Marshall government to unlock gas supplies before agreeing to construct an interconnector between NSW and South Australia (ABC News 2021). In the event the Marshall government maintained the broad scope and trajectory of South Australia's energy transition, choosing not to reverse its gains (which had proven popular with voters). With this bipartisanship, South Australia effectively announced the end of the so-called Climate Wars in Australia and anticipated the later conversion to renewable energy across other conservative-led State governments, notably in NSW, which adopted a wholesale Electricity Infrastructure Roadmap in 2020 (EnergyCo 2024).

The Marshall government was ejected by the electorate in 2022 after one term in office. The incoming Labor government was determined to resurrect the longstanding goal of decarbonising the state. With 14 utility-sized wind farms, an expanding array of solar farms, and a bank of batteries enabling the firming of supply, South Australia was put back on track of being able to rely on renewable energy. In 2023, renewable energy accounted for 75% of all energy consumed in South Australia, with wind power accounting for 48% of energy, rooftop solar 20%, and utility solar 6% (OpenNEM 2023). There was continued growth in the uptake of rooftop solar panels and increasing investment in utility solar farms, with some mega-sized projects securing debt finance from the Federal-level Clean Energy Finance Corporation and the ARENA (see ARENA 2021). There is now an ever-expanding number of hours during the day when demand is met wholly from renewable sources, and increasing storage capacity is extending these hours.

South Australia is certainly regarded as a renewable energy success story. The development of the utility-scale wind and solar farms, combined with battery storage, has provided a model for the rest of Australia. With the completion of two additional connectors, one to Victoria and a second to NSW, worries about security of supply are diminishing. The success of the transition is also evident in the scale of private investment in expanding generating capacity. As noted, the development has been made possible because of the extent to which international energy corporations have invested in the South Australian project, with the nexus of state capital injecting new momentum into the new renewable energy-based regime of accumulation.

There are downsides. As we discuss in Chapter 4, large-scale renewable energy has widescale impacts on society and the environment. Policies may be highly contested and force new agendas onto the table. Neoliberalised energy imposes its own priorities, against local interests. Marketisation, speculation, and the wider profit take can force up energy prices at a time when renewable energy technology offers the possibility of energy with zero marginal costs. The paradox of relying on private investors to solve the climate crisis can unravel even the most sophisticated 'market' apparatus.

Certainly, the transition has come at some cost to South Australian residents. South Australians have confronted some of the most expensive electricity in Australia, if not the world, and while the high rates of renewably generated electricity have brought wholesale prices down, this has not been passed on to consumers (Vorrath 2022). More noteworthy is the distributional effect of extending access to household solar and energy storage. The government envisages a network of potentially 50,000 solar and Tesla Powerwall home battery systems across South Australia contributing over 14% of the state's power (Government of South Australia 2023c). Yet, the promise of households reaping a return on their investment has been eroded with the decline in the value of feed-in tariffs. Remarkably enough, energy retailers such as AGL are now able to switch off household access to the grid in order to 'stabilise' energy prices (and maintain their rate of return). New questions now arise as South Australia plans for 500% renewables with a wave of anticipated 'downstream' industrial development, fuelled by hydrogen (Government of South Australia 2023b).

Conclusions

Government responses to climate crises and emissions reduction imperatives are directly linked to the politics of new renewable energy regions. Sub-national actors are able to plan and approve wind parks and solar farms, framing the increase in renewable energy in terms of regional co-benefits. In doing so, they incorporate renewable energy into regional development narratives as an opportunity, rather than as a burden. Regional policy is thereby aligned with green economy aspirations and may be deepened through public engagement in renewables, including through efforts to socialise benefits and ownership. That said, the regional embrace of renewable energy is highly uneven, itself reflecting the unevenness of fossil fuel-based development. Some regions pursue renewable energy as a means of fostering socio-economic development, others are disinterested or latecomers; and some, particularly those entrenched in fossil fuels, resist the transition.

Regional renewable narratives may be rejected, they may also be viewed as a passing preoccupation, overtaken by other concerns or by the logic of climate change. The regional renewable story is unfolding against a backdrop of radical uncertainty. There are uncertainties not least with the vagaries of climate policy internationally and in national politics and also with the high-risk dynamics of financialised energy. Renewable energy companies are able to exert increasing influence over policy and outcomes, compounding wider uncertainties over the technology itself and its social organisation, from utility-scale installations to households. As a governance trend, renewable energy regionalism is highly contingent, reflecting the wider currents of climate change and energy transition, which by definition are not preordained.

The federalised energy systems of India, Australia, and Germany have encouraged the emergence of distinctive region-level energy regimes, each with its own unique history, developmental trajectory, and electricity mix. The concept of renewable energy regionalism investigates the role of these regional agencies and authorities in enabling, steering, and delivering the regional energy transition. As the energy transition accelerates, we argue that regional bodies gain a pivotal role in regulating the process and enabling its outcomes. Regional bodies play a central role in leveraging legitimacy for renewables, enabling new forms of democratic involvement and social ownership. The multi-level character of renewables development can bolster the region's centrality, as regional bodies vie for influence with each other, with (and against) the central state, and with corporates. The success of regional strategy in turn depends on social legitimacy. The extent of region-level public engagement, driven to varying degrees by political leadership, bureaucratic entrepreneurship, regulatory development, and civil society action, forms a key aspect to understanding the changing scope for renewable energy legitimacy.

The framework developed in Chapter 1 suggests a focus on legitimising renewable energy, and specifically on the process of legitimising appropriation, accumulation, and regulation. The historical development of renewable energy in the three case study countries, as outlined here, demonstrates how these three dimensions have shaped outcomes on the ground. The process of establishing corporate-owned large-scale renewable energy in the three national contexts has directly raised key issues of ecological appropriation, from land use change, to impacts of technology, to questions of location. The issue of who gains and who loses from these installations, in terms of distributional justice, has become a central focus of disputation, centring on such issues as land pricing and rental, access to jobs and royalties, and potential for social ownership. Finally, the overarching question of public engagement and involvement in the regulatory process and in the wider process of defining a regional identity in terms of renewable development has created a powerful regional political contest.

Chapters 3 to 5 shift our focus from theoretical frameworks and policy regimens to lived experiences in the case study regions. The chapters provide an up-close examination of the socio-political dynamics, cultural shifts, and economic repercussions brought about by the transition to renewable energy, shedding light on the lived realities amidst the larger landscape of global environmental shifts and energy reforms. Their shared ethnographic approach seeks insights into these dynamics over several years, beyond the scope of a more snapshot approach. The focus on public engagement centres on the social agency of participants, across landholders, civil society, and local government, highlighting important lessons in the social legitimacy of the energy transition.

We thereby seek to offer a granular exploration of how the energy transition is woven into the fabric of daily life, grounded in the specificities of each region and the perspectives of local communities, policymakers, and key stakeholders. Karnataka's sun-drenched landscapes, South Australia's semi-arid expanses, and Brandenburg's wind-swept plains set the context for surprisingly similar narratives across hope, disenchantment, contestation, and regulation. With this, legitimacy becomes a key aspect of the transition process.

References

ABC News (2005). 'Rann calls for urgent climate change meeting', ABC News, 5 January 2005.

ABC News (2019). 'Port Augusta solar thermal power plant scrapped after failing to secure finance', ABC News, 4 April 2019.

ABC News (2021). 'Scott Morrison reveals $1bn energy deal with South Australian government', ABC News, 18 April 2021.

AEMO (2016) South Australian Renewable Energy Report, Adelaide: Australian Energy Market Operator.

AEMO (2022). *AEMO Insights: South Australia*, Adelaide: AEMO.

Agrawal, S., Mani, S., Jain, A. & Ganesan, K. (2020). *State of Electricity Access in India: Insights from the India Residential Energy consumption Survey (IRES) 2020*, New Delhi: Council on Energy, Environment and Water.

Appunn, K. (2023). 'Coal in Germany', Clean Energy Wire, 18 January 2023.

ARENA (2017) 'Australia's rooftop solar produces enough energy to power this city', ARENAwire, 31 August 2017.

ARENA (2021). '$100 million to build large scale battery storage', Media Release, Australian Renewable Energy Agency, 19 December 2021.

Barman, A. & Pathak, K. (2022). 'CPPIB infuses $400m more in ReNew Power', Economic Times, 12 November.

BDEW (2024) *Karten Daten Fakten zur Energiewirtschaft, Federal Association of the Energy and Water Industries*, Berlin: BDEW.

Bedi, H. (2022). Solar power for some? Energy transition injustices in Kerala, India. *Environment and Planning E: Nature and Space*, 5(3), 1146–1163.

BEE (2013). *Stellungnahme zum Energieteil des Koalitionsvertrages, Federal Association of Renewable Energy*, Berlin: BEE.

Bhattacharyya, S., Palit, D., Sarangi, G., Srivastava, V. & Sharma, P. (2019). Solar PV mini-grids versus large-scale embedded PV generation: a case study of Uttar Pradesh (India). *Energy Policy*, 128, 36–44.

BMWK (2014). *Germany Is a Pioneer in the Expansion of Renewable Energies*. Berlin: Federal Ministry of Economics and Climate Action.

Boute, A. (2013). Renewable energy federalism in Russia: regions as new actors for the promotion of clean energy. *Journal of Environmental Law*, 25(2), 261–291.

Breusch, J. (2005). 'Rann calls for climate commitment', The Australian Financial Review, 5 January 2005.

Bridge, G. & Gailing, L. (2020). New energy spaces: towards a geographical political economy of energy transition. *Economy and Space*, 52(6), 1037–1050.

Buckley, T. & Shah, K. (2018). *Karnataka's Electricity Sector Transformation*, Delhi: The Institute for Energy Economics and Financial Analysis.

Bundesgesetzblatt (2000). *Gesetz für den Vorrang Erneuerbarer Energien (Renewable Energy Act)*, Berlin: Bundesgesetzblatt (Federal Law Gazette).

Bundesgesetzblatt (2011). *Dreizehntes Gesetz zur Änderung des Atomgesetzes (Laws amending the Atomic Energy Act)*, Berlin: Bundesgesetzblatt (Federal Law Gazette).

Bundesnetzagentur (2024a). *Statistiken ausgewählter erneuerbarer Energieträger zur Stromerzeugung (Selected statistics: Renewable energy sources for electricity generation)*, Berlin: Bundesgesetzblatt (Federal Law Gazette).

Bundesnetzagentur (2024b). *Netzentgelte (Regulatory Decisions)*, Berlin: Bundesnetzagentur (Federal Network Agency).

Bundesregierung (2011). 'Energiewende, Die Gesetze' (Energy Transition: The Laws), German Federal Government, 5 August 2011.

Bundesregierung (2023) *Energieversorgung in Deutschland (Energy Production in Germany)*, Berlin: Statistisches Bundesamt.

Burroni, L. (2014). Competitive regionalism and the territorial governance of uncertainty. *European Review of Labour and Research*, 20(1), 83–97.

Byzio, A., Mautz, R. & W. Rosenbaum (2008). *Towards Energy Transition: The Development of Electricity Production from Renewable Energies Germany*, Göttingen: Sociological Research Institute of Göttingen.

Central Electricity Authority (2023). *National Power Portal*, Delhi: Ministry of Power, Government of India.

Chapman, A., McLellan, B. & Tesuka, T. (2016). Residential solar PV policy: an analysis of impacts, successes and failures in the Australian case. *Renewable Energy*, 86, 1265–1279.
Chye, M. & Chew, C. (2023). 'India's power output grows at fastest pace in 33 years, fuelled by coal', Reuters, 6 April 2023.
DCCEWE (2023). *Australian Energy Update*, Canberra: Department of Climate Change, Energy, Water and Environment.
De Laurentis, C. & Pearson, P. (2021). Policy-relevant insights for regional. Renewable energy deployment. *Energy, Sustainability and Society*, 11(1), 1–21.
Deepika K. (2022). 'Karnataka emerges as top seller of renewable energy under Green Term Ahead Market', The Hindu, 3 April 2022.
Deutschlandfunk (2021). 'Germany achieves its 2020 climate targets – thanks to Corona', Deutschlandfunk, 5 March 2021.
Deutsche Windguard (2019). *Status des Windenergieausbaus an Land in Deutschland 2019 (Wind Energy Statistics: Year 2019)*, Varel: DW.
Deutsche Windguard (2023). *Status des Windenergieausbaus an Land in Deutschland 2019 (Wind Energy Statistics: Year 2013)*, Varel: DW.
Doczi, S. (2021). 'International Context for RE Integration in Karnataka', International Energy Agency, Presentation, 19 January 2021.
Elavarasan, R. et al. (2020). A comprehensive review on renewable energy development, challenges, and policies of leading Indian states with an international perspective. *IEEE Access*, 8, 74432–74457.
EnergyCo (2024). *The Electricity Infrastructure Roadmap*, Sydney: EnergyCo.
EnergyMap (2016) *Energieregionen (Energy Map)*, Berlin: DGS.
European Commission (2022). *REPowerEU Plan*, Brussels: European Commission.
Fabra, N., Matthes, F. C., Newbery, D., Colombier M., Mathieu, M. & Rüdinger, R. (2015). *The Energy Transition in Europe: Initial Lessons from Germany, the UK and France: Towards a Low Carbon European Power Sector*, Brussels: Centre on Regulation in Europe.
Fedorowytsch, T. (2015). 'States can take lead over climate change, South Australian Premier Jay Weatherill says', ABC News, 4 December 2015.
Förster, F. (1968). *Senftenberger Revier 1890–1914. On the history of the Lower Lusatians Lignite Industry from the Case of the Socialist Act to the Outbreak of the First World War*, Bautzen: Domowina-Verlag.
Gailing, L., Hüesker, F., Kern, K., & Röhring, A. (2013). The spatial design of the energy transition, between centrality and decentralization, Working Paper 51, Leibniz: Institute for Research on Society and Space.
Gibbs, D. & O'Neill, K. (2017). Future green economies and regional development: a research agenda. *Regional Studies*, 51(1), 161–173.
Goodman, J. & Rosewarne, S. (2015). Slowing uranium in Australia: lessons for 'urgent transition' beyond coal, gas and oil. In T. Princen, J. Manno & P. Martin, eds., *Ending the Fossil Fuel Era*, Cambridge: MIT Press, pp. 193–222.
Government of Karnataka (2022). 'Notification: Karnataka Renewable Energy Policy 2022–2027', Government of Karnataka, 16 April 2022.
Government of South Australia (2023a). *Leading the Green Economy*, Adelaide: GoSA.
Government of South Australia (2023b). *Hydrogen and Renewable Energy Act*, Adelaide: GoSA.
Government of South Australia (2023c). *South Australia's Virtual Power Plant*, Adelaide: GoSA.
Gulia, J., Sharma, P. & Garg, V. (2022). *Renewable Energy Tenders Issuance in India: Not in Tandem with Government Targets*, Delhi: The Institute for Economics and Financial Analysis.

Gurucharan, G. (2022). Karnataka's energy sector: advancing second generation distribution reforms, Report of the Committee on Restructuring the Power Sector, Bangalore: Government of Karnataka.

Hancock, K. (2024). Introduction to the special issue on energy regionalism. *Review of Policy Research*, 41, 282–289.

Helgenberger, S. & Jänicke, M. (2017). Connecting opportunities with interests in the new energy world of renewables. Working Paper, Potsdam: IASS.

Hettchen, S. (2020) *Das Brandenbuch: Ein Land in Stichworten (The Brand Book: A Country in Keywords).* Potsdam: Brandenburg State Agency for Political Education.

Hillebrand, R. (2013) Climate protection, energy security, and Germany's policy of ecological modernisation. *Environmental Politics*, 22(4), 642–668.

Hoberg, G. (2021). *The Resistance Dilemma: Place-Based Movements and the Climate Crisis*, Cambridge: MIT Press.

Holz, F., Hirschhausen, C., Sogalla, R., Barner, L., Steigerwald, B. & Kemfert, C. (2023). *Deutschlands Gasversorgung ein Jahr nach russischem Angriff auf Ukraine gesichert (Germany's Gas Supply Secured One Year after Russian Attack on Ukraine)*, Berlin: DIW (German Insttute for Economic Research).

IBEF (2021). 'India Renewable Energy Industry Report', India Brand Equity Foundation (IBEF), June 2021.

IEA (2021). *Renewables Integration in India*, Paris: International Energy Agency.

IEA (2022). *Installed Capacity in India, 2000–2020, and Projections up to 2040 in the Stated Policies Scenario*, Paris: International Energy Agency.

Jehling, M., Hitzeroth, M. & Breuckner, M. (2019). Applying institutional theory to the analysis of energy transitions: from local agency to multi-scale configurations in Australia and Germany. *Energy Research & Social Science*, 53, 110–120.

Jenniches, S. (2018). Assessing the regional economic impacts of renewable energy sources – a literature review. *Renewable and Sustainable Energy Reviews*, 93, 35–51.

Johnson, C. & VanDeer, S. (2021). Energy regionalisms in theory and practice. *Review of Policy Research*, 41(2), 290–309.

Jordaan, S., Davidson, A., Nazari, J. & Herremans, I. (2019). The dynamics of advancing climate policy in federal political systems. *Environment Policy and Governance*, 29(3), 220–234.

Jörgensen, K., Mishra, A. & Sarangi, G. (2015). Multilevel climate governance in India: the role of the states in climate action planning and renewable energies. *Journal of Integrative Environmental Sciences*, 12(4), 267–283.

Kale, S. (2014). *Electrifying India: Regional Political Economies of Development*, Stanford: Stanford University Press.

Katakey, R. & Chakraborty, D. (2014). 'Modi to use solar to bring power to every home by 2019', Bloomberg, 19 May 2014.

Kelton, G. (2010). 'Rann heads overseas for climate talks', The Advertiser, 1 December 2010.

Kenning, T. (2017). 'India's SECI financially strengthened under government and RBI agreement', PV-Tech, 16 February 2017.

Koalitionsvertrag (2021). *'Koalitionsvertrag 2021–2025 zwischen der SPD, Bündnis 90/ Die Grünen und FDP' (Coalition Agreement)*, Berlin: SPD.

KREDL (2023a) *Procedure for Allotment of Solar Park Projects*, Bangalore: Karnataka Renewable Development Limited.

KREDL (2023b): *Progress of Solar Power Projects in Karnataka*, Bangalore: Karnataka Renewable Development Limited.

KREDL (2023c). *Details of Year-Wise Commissioned Projects of Renewable Category in MW*, Bangalore: Karnataka Renewable Development Limited.

KREDL (2023d). *Commissioned Wind Power Projects*, Bangalore: Karnataka Renewable Development Limited.

KREDL (2023e). *District Wise RE Power Projects Commissioned*, Bangalore: Karnataka Renewable Development Limited.

Kumar, A. & Thapar, S. (2017). *Addressing Land Issues for Utility Scale Renewable Energy Deployment in India*, Delhi: Teri School of Advanced Studies.

Lakhanpal, S. (2019). Contesting renewable energy in the global south: a case-study of local opposition to a wind power project in the Western Ghats of India. *Environmental Development*, 30, 51–60.

Leiren, M. & Reimer, I. (2018). Historical institutionalist perspective on the shift from feed-in tariffs towards auctioning in German renewable energy policy. *Energy Research and Social Science*, 43, 33–40.

Lexikon (2015). *Lexikon der Nachhaltigkeit: Koalitionsvertrag 1998–2002 (Lexicon of Sustainability: Coalition Agreement 1998–2002)*, Koblenz: Kompagnon Publishing.

Löschel, A., Grimm, V., Lenz, B. & Staiß, F. (2021). *Statement on the Eighth Monitoring Report by the Federal Government for the Reporting Years 2018 and 2019, (Expert Commission for the Monitoring Process 'The Energy of the Future')*, Berlin: Expert Commission.

Lutz, L., Fischer, L., Newig, J. & D. Lang (2017). Driving factors for the regional implementation of renewable energy: a multiple case study on the German energy transition. *Energy Policy*, 105, 136–147.

Mattes, J., Huber, A. & Koehrsen, J. (2015). Energy transitions in small-scale regions. *Energy Policy*, 78, 255–264.

Meya, J., Neetzow, P., Neubauer, L. & Pechan, A. (2017). Die Menge macht's? Das EEG 2017 und die Folgen für die deutsche Energiewende (The quantity does it? The EEG 2017, consequences for the German energy transition). *Energiewirtschaftliche Tagesfragen*, 66(11), 34–37.

Michael, S., Lolla, A. & Gard, V. (2023). *Indian States' Electricity Transition (SET): Evaluating the Preparedness of 16 Indian States to Walk the Electricity Transition*, Delhi: The Institute for Energy Economics and Financial Analysis.

Ministry of New and Renewable Energy (2023a). *Solar Trajectory 2022*, MNRE: Delhi.

Ministry of New and Renewable Energy (2023b). *Physical Achievements*, MNRE: Delhi.

Ministry of New and Renewable Energy (2024a). 'Cabinet approves PM-Surya Ghar: Muft Bijli Yojana for installing rooftop solar in One Crore households', 29 February, MNRE: Delhi.

Ministry of New and Renewable Energy (2024b). *Rewnewable Energy Statistics 2023–24*, Ministry of New and Renewable Energy: Delhi.

MLUK (2024) *Renewable Energy*, Potsdam: Ministrylture, Environment and Climate Protection, State of Brandenburg.

Moezzi, M., Janda, K. & Rotmann, S. (2017). Using stories, narratives, and storytelling in energy and climate change research, *Energy Research and Social Science*, 31, 1–10.

Mookerjea, S. (2019). Renewable energy transition under multiple colonialisms: passive revolution, fascism redux and utopian praxes. *Cultural Studies*, 33(3), 570–593.

Moss, T., Becker S. & Naumann, M. (2015). Whose energy transition is it, anyway? Organisation and ownership of the Energiewende in villages, cities and regions. *Local Environment*, 20(12), 1547–1563.

Müller, K. (2017). Heimat, Kohle, Umwelt: Argumente im Protest und der Befürwortung von Braunkohleförderung in der Lausitz (Home, coal, environment: protest and advocacy over lignite mining in Lusatia). *Zeitschrift für Umweltpolitik und Umweltrecht*, 3, 213–228.

Müller, K. & Pampus, M. (2023). The solar rush: invisible land grabbing in East Germany. *International Journal of Sustainable Energy*, 42(1), 1264–1277.

MWAE (2022). Energy Strategy 2040, Ministry of Economics, Labor and Energy, State of Brandenburg.

MWAE (2024). Wind Energy, Ministry of Economics, Labor and Energy, State of Brandenburg.

O'Brien, R. (2009). 'Rann puts SA at forefront of renewable push', Government of South Australia Media Release, 17 December 2009.

Ohlhorst, D. (2015). Germany's energy transition policy between national targets and decentralized responsibilities. *Journal of Integrative Environmental Sciences*, 12(4), 303–322.

OpenNEM (2023). *Energy – South Australia*, Adelaide: OpenNEM.

Overwien, P. & Groenewald, U. (2015) *Much Ado about Wind. Current Challenges for Regional Planning in Brandenburg*, Berlin: Federal Institute for Research on Building, Urban Affairs and Spatial Development.

Parkinson, G. (2011). 'Rann signs off with a clean vision', InvestSmart, 11 October 2011.

Parkinson, G. (2018). 'Speed of Australia's energy transition hostage to Marshall Law', Renew Economy, 19 March 2018.

Parkinson, G. (2023) 'Rooftop solar meets all of South Australia demand in major new milestone', Renew Economy, 24 September 2023.

Pinkera, A., Argüelles, L., Fischerd, A. & Becker, S. (2020). Between straitjacket and possibility: energy initiatives and the politics of regulation. *Geoforum*, 113, 14–25.

Pohl, A. (2023a). Energiemarkt Deutschland: Liberalisierung, Akteure & Preise', Energiemarie, 21 March 2023.

Pohl, A. (2023b). 'So funktioniert der Stromhandel Deutschland', Energiemarie, 30 October 2023.

Rann, M. (2020). 'Subnational governments hold the key to climate change', Financial Times 4 February 2020.

Rosewarne, S. (2022). *Contested Energy Futures: Capturing the Renewable Energy Surge in Australia*, London: Palgrave Macmillan.

SECI (2010). *Jawaharlal Nehru National Solar Mission: Towards Building Solar India*, Delhi: Solar Energy Corporation of India.

Shokrgozar, S., & Girard, B. (2024). 'The companies are powerful, people are weak': India's solar energy ambitions and the legitimation of dispossession in Rajasthan. *Journal of Political Ecology*, 31(1), 48–66.

Sijm, J. (2002). *The Performance of Feed-in Tariffs to Promote Renewable Electricity in European Countries*, Petten: Energy Research Centre of the Netherlands.

Simshauser, P. (2021). Lessons from Australia's national electricity market 1998–2018: the strengths and weaknesses of the reform experience. In J. Glachant, et al. eds., *Handbook on Electricity Markets*, London: Edward Elgar, pp. 242–286.

Singh, D. (2022). 'This is all waste': emptying, cleaning and clearing land for renewable energy dispossession in borderland India. *Contemporary South Asia*, 30(3), 402–419.

Slezak, M. (2008). 'All in on renewable energy: Jay Weatherill to ramp up SA target to 75%', The Guardian, 21 February 2008.

Smyth, J. (2017). 'Elon Musk makes bold bet to end Australia power crisis', Financial Times, 13 March 2017.

Späth, P. & Rohracher, H. (2010). Energy regions: the transformative power of regional discourses on socio-technical futures. *Research Policy*, 39(4), 449–458.

Spiegel (2011) 'SPD-Vordenker verlangt neuen Atomkonsens', Der Spiegel, 14 March 2011.

Statista (2024) Brown coal production in Germany from 1990 to 2022, Berlin: Expert Commission for the Monitoring Process.

Statistisches Bundesamt (2022). *Energy production*, Berlin: GoG.

Statistisches Bundesamt (2024a) *Energieabhaengigkeit der EU 2022 bei 63%*, Berlin: GoG.

Statistisches Bundesamt (2024b) *Photovoltaik: Deutschland größter Erzeuger in der EU*, Berlin: GoG.

Statistisches Bundesamt (2024c) *Energieerzeugung*, Berlin: GoG.

Stokes, D. (2022). Renewable energy federalism. *Minnesota Law Review*, 206, 1757–1815.

The Sydney Morning Herald (2007). 'Australia missed renewables chance: Rann', Editorial, 25 September 2007.

Umweltbundesamt (2022). *Renewable Energy in Germany 2021*, Berlin: Federal Environment Agency.

Umweltbundesamt (2023a). *Greenhouse Gas Emissions*, Berlin: Federal Environment Agency.

Umweltbundesamt (2023b). *Energy Projection Report*, Berlin: Federal Environment Agency.

Vorrath, S. (2017). 'Weatherill tears strips off Frydenberg for renewables "trash talk"'. Renew Economy, 16 March 2017.

Vorrath, S. (2022). 'SA renewables surge bringing down energy prices, but consumers miss out', Renew Economy, 11 May 2022.

Weller, S. (2018). Globalisation, marketisation and the transformation of Australia electricity sector. *Australian Geographer*, 49(3), 439–453.

Wettengel, J. (2020). 'A (very) brief timeline of Germany's Energiewende', Clean Energy Wire Factsheet, 15 May 2020.

Windbranche (2023a). *German Wind Energy Market*, Berlin: Industry Wind Energy Portal.

Windbranche (2023b). *Brandenburg: Wind Energy Expansion*, Berlin: Industry Wind Energy Portal.

Yenneti, K., Day, R. & Golubchikov, O. (2016). Spatial justice and the land politics of renewables: dispossessing vulnerable communities through solar energy mega-projects. *Geoforum*, 76, 90–99.

3

Karnataka, India

We were standing in the forecourt of a small house in a South Indian village near the town of Pavagada. Around us were a number of humble dwellings and dusty narrow streets with cows and goats. We wanted to talk to the women in that area, which was near the Pavagada Solar Park, about how this enterprise affected them. The women called us into the forecourt of one of those houses, and we began to chat.

At that point, several men turned up. Many were angry. They created a huge confrontation. There was a lot of shouting and yelling. It looked like the situation could get ugly and so our interpreter, accompanying our group of five or six female and male researchers from Australia and India, said that maybe we should just leave. We said to the women, 'Look we will come back another time'.

And the women really astonished us. They literally grabbed our hands and said, 'No, we want you to stay. We want to tell you our story'. They dragged us into a room off the forecourt and locked the door and made us sit down in there.

A woman sitting next to me, an older woman, was obviously very agitated by what had happened. She kept hitting me [Devleena Ghosh] painfully on the knee and saying, 'You must do something. You must do something. You have the ear of the people in authority. You have to tell them how bad things are for us. You must do something to change our lives'. These women explained to us that nearly everyone in their village had lost their livelihoods and was now at risk of severe poverty. And all this was happening because of a new solar farm in the area.

> This area could be really thriving. There was a plan for the solar farm for the region. It was a great idea. If it had been done properly, thousands of lives would have been changed for the better. But it is not working. There is a significant part of the community missing out on the benefits of this plan and they are suffering because of it.

Introduction

This account describes how one of the authors of this book experienced the visit we paid to the people living with and next to the Pavagada Solar Park (53 km^2 and with 2,050 MW of generating capacity) in the Indian state of Karnataka. It flags the issues and emic perspectives of people living in the area. This chapter focuses on this solar park, and other wind power developments, to describe and analyse the entry of renewable energies into India. Since 2010, Karnataka renewable energy development has centred almost exclusively on large-scale corporate-led renewable energy projects. This chapter focuses on this recent phase in the development of renewable energy in the state, highlighting its drivers, beneficiaries, and losers, and how it may develop into the future. A key dimension is the institutional learning process that proceeds at the state level, in terms of the policy vehicles used to advance energy transition and their associated lock-ins. State authorities, corporate players, and affected communities are all drawn into a process of engagement with renewable energy and its development. Each is actively and proactively generating new strategies in the context of rapidly developing technological possibilities, particularly for up-scaled projects on ‘greenfield’ sites, growing demand from industrial and commercial users, and increasing concern about the environmental and climate impacts of existing coal-fired power stations. That process of strategic engagement reflects the uneven and unequal logic of the transition, and in particular the social stratification that it reflects and exacerbates. The discourses promoting renewable energy are focused not only on their neutral impact on climate change but also on scale – big projects, big energy production, big benefits, and for the companies involved, big profits.

The account here first outlines the state-level institutional context, particularly in terms of the use of empowered ‘special purpose’ agencies dedicated to advancing renewable energy in the context of an otherwise faltering market-based energy system. It then discusses wind power projects and the move to larger-scale corporate-led installations, focusing on an early relatively large-scale corporate project in Shimoga, central Karnataka, which was constructed on state-owned forest land. It then analyses policy for large-scale solar projects, adopted in 2010, and focuses on the largest of these, the 2 GW Pavagada solar park, located in the east of Karnataka, in the Tumkur region. Finally, the chapter concludes by exploring the commitment of Karnataka state to a 5 GW ‘hybrid’ installation with wind, solar, and storage centred on the Shimoga region and a doubling of the Pavagada solar park, which we discuss in light of lessons from the legitimacy of policy tools and installations emanating from struggles for social rights within the dynamics of accumulation, appropriation, and regulation discussed in the chapter. The chapter is based on national and state-level fieldwork in Delhi and Bangalore and local fieldwork conducted in 2017–2019

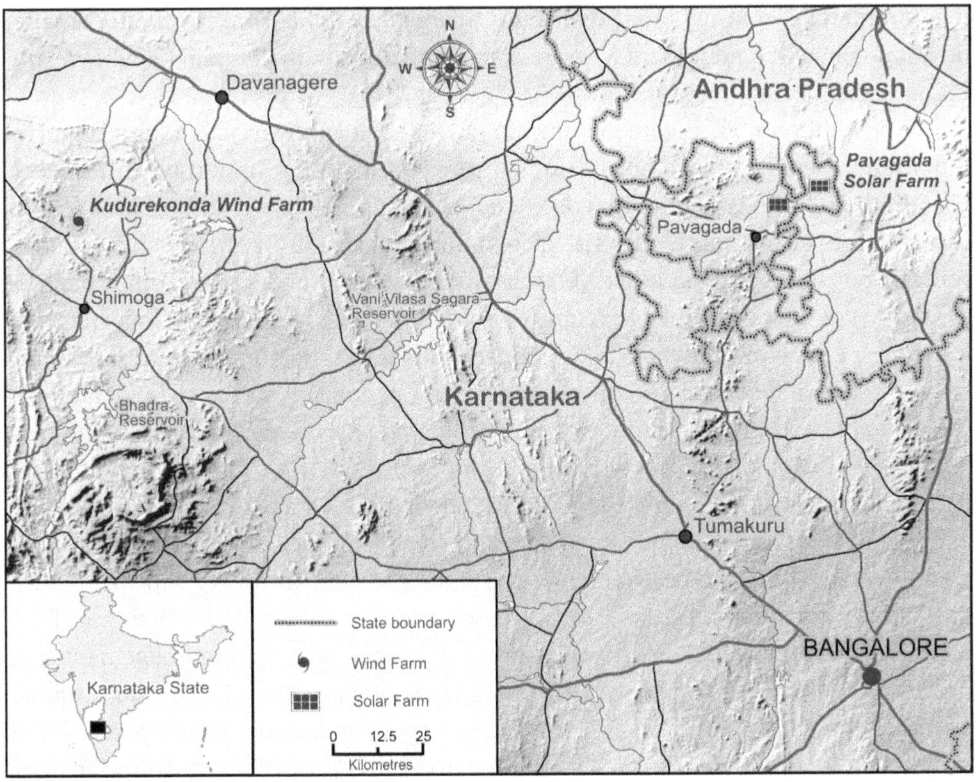

Map 3.1 Karnataka selected renewable projects (*Source*: Olivier Rey-Lescure 2024).

for Pavagada and in 2018–2022 for Shimoga (see Map 3.1). Insights are drawn from state agencies, corporate players, think tanks, NGOs, and representatives from local communities affected by installations.

Institutionalising Renewable Energy

As outlined in Chapter 2, the upsurge of renewable energy in Karnataka reflects a confluence of factors: favourable wind and solar resources, the need to overcome reliance on coal imports for coal-fired power, rapidly growing energy demand from electrification, a desire to attract investment capital, the need for cheaper power generation to reduce costs for loss-making state-owned electricity distribution companies, and the consolidation of export markets for electricity in neighbouring states and countries. These factors have underpinned a process of institutional innovation centred on facilitating collaboration with large-scale private investment capital. They have also opened up spaces for contestation over other rights claims, specifically from the landless population who have been excluded from the commercial contracts that distribute financial benefit from renewable energy development and landowners who are also conflicted about unequal benefits resulting in social destabilisation.

The institutional arrangements for the power sector in Karnataka largely mirror the arrangements in place across Indian states, with some important differences. Electricity in India is a 'dual mandate' sector, under the shared authority of federal and state governments. Under the Constitution, electricity is listed as a 'concurrent' responsibility of state and federal governments. Land, on the other hand, is a solely state responsibility (except for forest land, which is also concurrent and a major site of wind power). These constitutional delineations require joint work between state and central government agencies on securing access to land, facilitating grid connections, and arranging public financing. Federal initiatives in renewable energy, such as the Solar Energy Corporation of India (SECI), have played a critical enabling role, for instance, in requiring the establishment of renewable energy development agencies in all states to work with the central government and also in establishing a renewable energy funding stream for energy infrastructure, particularly for transmission. Funds are sourced from federal revenue, and from a range of international financial institutions, especially development banks, to underwrite up-front fixed costs. Federal agencies then establish 'special purpose' agencies with state counterparts to advance specific large-scale projects. These projects strike up Power Purchase Agreements (PPAs) with state distribution companies (DISCOMs; and also with the central Federal-run Indian Energy Exchange platform) in tender processes managed by special purpose state authorities.

The Karnataka government's special purpose agency for renewable energy is KREDL, short for Karnataka Renewable Energy Development Limited. While wholly owned by the state government, KREDL has been established as a special purpose company with a commercial orientation. Government officials are aware that local communities are inclined to trust quasi-governmental bodies and KREDL's involvement has been essential to the process of securing leasehold land for renewable energy projects. We met with the assistant general manager for wind energy as well as with the managing director and head of KREDL in their newly erected building in the south of Bengaluru. The compound featured the obligatory security officer and a bright and shiny entrance hall, which comes with a thematically fitting installation of vertical plants and solar panels. The building and its compound also feature rooftop solar and charging stations for e-vehicles. The managing director and his team not only pointed out those facilities but also gave us a presentation of renewable energy development in Karnataka and KREDL's contribution to it.

The presentation not only outlined the significant increase of renewable energy installed capacity in Karnataka but also gave the managing director the opportunity to point out the favourable policy environment that allowed for this development. Every leap in renewable energy capacity in the state, including but also beyond the installation of Pavagada solar park, was attributed to the advances and innovations in the policy framework. KREDL's managing director pointed out initiatives across the renewable energy supply chain, including an incubation centre for renewable energy-related start-ups and innovations, a skill development centre to provide the required qualified workforce in the domain, and a professional internship training centre for students of engineering colleges. A core aspect of their policy work, the managing director and his staff stressed, is the so-called one window clearance. 'We try to make it as easy as possible for investors here in Karnataka', ensuring land procurement in 10–15 days and a clearance of all permits related to the installation of renewable energy within 15 days or, for more complex cases, 60 days, with the investor needing to approach only one authority. KREDL assures developers that they will secure all land use approvals and regulations on their behalf with relevant state and federal authorities. This contributed to a view from developers, according to an official we interviewed from the local office of the World Research Institute, that 'it's almost that if I pay the fee I will complete the project'.

The central role of KREDL is therefore to facilitate approvals, remove planning impediments, and minimise establishment costs. KREDL, as a commercially oriented entity, in turn generates its own assessment and processing fee income for 'infrastructure development facilitation', as well as income from land sub-leases and the sale of power from small and demonstration generation installations. Partly as a result, KREDL has seen its income from operations increase from 245 million Rs. in 2010 to 1.03 billion Rs. in 2022 (KREDL 2010; 2022). But more importantly, the KREDL model is to support profitable accumulation of renewable energy capital by intermediating long-term supply contracts that lock in future sales and prices, secure the appropriation of land needed for large-scale wind and solar across complex landholding relations, and smooth conflicts that emerge over the regulation process by bypassing processes that are most contentious in our other cases, such as environmental and planning approvals.

In addition to facilitating independent renewable energy projects, KREDL has sought to establish its own large solar parks, where private operators are able to operate on a 'plug and play' basis. In such cases, KREDL has set up second-tier agencies jointly with the federal government, specifically to acquire land, install infrastructure, and host private solar or wind developers bidding for PPAs. This is the model that was developed for the Pavagada solar project, with the creation of the Karnataka Solar Power Development Corporation Limited (KSPDCL), a joint venture between KREDL and the centrally controlled SECI Limited, which oversees India's National Solar Mission for the Ministry of New and Renewable

Energy (MNRE). KSPDCL was established specifically for the Pavagada solar park as a state-owned intermediary designed to overcome the difficulties that private developers face in acquiring land and establishing infrastructure.

KREDL's role as an intermediary rests, among other things, on providing data to investors, such as complete wind potential analysis on a very localised *taluk* level. 'After looking at the data, the companies need to go in with a consultant or anyone to find out the most suitable spots in the *taluk*', the managing director explains. On the other side, towards landholders and regular local citizens, KREDL as a state institution is also a trusted party. 'Our incentives for what we as KREDL do, are taxes, jobs and CSR [corporate social responsibility]', says KREDL's managing director. In other words, KREDL as a state-owned company acts to achieve state policy goals by securing profitability for private investors. Despite its commercial orientation, the position of KREDL as a public agency is crucial to gaining community trust in land acquisition and leasing arrangements. Landholders are more likely to put trust in authorities with public interest and agree to relinquish or lease lands for renewable energy through them.

Next to KREDL as a state agency, federal agencies have a particular role in facilitating access to land. First and foremost, and in contrast to our Australian and German cases, the Indian federal government has given blanket environmental approvals exempting renewables from environmental impact assessments and pollution control board approvals. The federal government also has a particular role in facilitating access to centrally owned forest lands, which were a major site of early wind power development in Karnataka. As will become evident in the Shimoga case study laid out later, the role of federal authorities in facilitating land for wind has been criticised for overlooking local communities more so than when the state government has a wider role. As a representative from the Indian Wind Power Association told us, unless there is a tribal connection that triggers other consultation processes, '*Panchayats* [local councils] have no say in forest lands'.

Federal agencies are also critical for securing finance for different aspects of renewable energy development. Federal agencies are able to leverage development finance for large-scale projects, especially for transmission infrastructure. As discussed in Chapter 2, SECI is the central federal government agency that facilitates the rollout of solar (SECI 2010). An official from SECI we spoke to described its role as cutting across 'Finance, connectivity, land – all these things we are facilitating'. They outlined that the overarching goal of SECI in facilitating these key components of renewable energy development is creating an attractive environment for international investors by de-risking investments in the sector: 'The investor must have the confidence if they are coming all the way to India'. In Karnataka, both SECI and KREDL have intermediate PPAs between generators and distributors, conducting auctions and guaranteeing payments. In this context the private generators can maintain low transaction and establishment costs, renting land and securing access to transmission networks via public intermediaries, then supplying energy to the distribution companies at the agreed rate.

Karnataka's embrace of renewable energy articulates with the ongoing financial difficulties of the distribution companies (generally known as DISCOMs in India but known as ESCOMs in Karnataka). DISCOMs are the companies that buy power wholesale and distribute it to consumers. As outlined in Chapter 2, Karnataka's DISCOMs, like DISCOMs across India, have become increasingly indebted and reliant on bailouts from state and federal governments. The financial issues facing Karnataka's DISCOMs stem from purchase costs increasing faster than revenue from sales of power. This is due to a combination of factors including the provision of free electricity for agricultural irrigation, other non-billed connections and high transmission losses, and exposure to high and spiking power purchase costs from thermal coal and hydropower (Buckley and Shah 2018; Garg and Sundararagavan 2018: 15; Gurucharan 2022: 16). The growth of renewable energy has dramatically reduced the costs of power purchased by DISCOMs for distribution to consumers while shielding other aspects of the electricity system from politically difficult reforms.

Securing lower cost power from renewable energy generators has been significant for Karnataka's DISCOMs because the prices DISCOMs can charge consumers are heavily regulated. When Karnataka separated generation from distribution in reforms implemented following the Electricity Act 2003, it maintained the pricing regulator, the Karnataka Electricity Regulatory Commission. The Commission is empowered to fix electricity prices on a year-by-year cycle. Distributors submit their accounts to the regulator and the electricity tariff is determined on the basis of these accounts and in response to extensive public hearings and submissions. The regulator publishes their decision, along with the distributors' accounts, determining tariffs for subsets of electricity consumers (KERC 2024). In Karnataka, electricity is subsidised for rural communities and for urban households below a certain rate of energy consumption. Commercial users and households with a higher rate of consumption pay more. The Karnataka state government provides a direct subsidy to meet the resulting shortfall in revenue, although tariffs overall are capped by the regulator. This system of price fixing reflects the politicisation of electricity, which has resulted in it being seen as a public necessity, a type of civic commons: pricing is in this respect subject to public interest considerations, with the state acting to ensure the continued functioning of the market for social ends, even in the context of the liberalisation of the economy, including energy systems. However, these restrictions on price have combined with other factors to create financial troubles for the distribution companies.

As with other states, in Karnataka, the distribution companies are the largest wholesale buyer of power in the Indian electricity market. Their PPAs with coal-fired power providers, which are mostly state-owned, are long-standing. Power purchase costs paid by distribution companies to generators have been inflated by the high cost of coal arising from shortages and the high fixed costs of operating thermal power stations. These costs have been pushed on to DISCOMs, which have contracts to pay the fixed operating costs of coal-fired power stations, even when their generation is not required. New coal-fired power is prohibitively costly, with most new coal power development proposals not commencing construction, forcing investment for new power into renewable energy. Yet distribution companies seek to source power from lowest-cost renewable energy generators to meet daytime power needs, but at the same time maintain coal-fired power supplies, even at a higher per unit rate, to cover for the dusk-to-dawn fall-off in solar power, and to compensate for the intermittency of wind energy.

This approach to renewables reflects the sentiment expressed to us by SECI, which said they were attempting to strike a fine balance between 'not disturbing what is there so far' while planning for a renewable future to meet demand in a market where 'there is no new thermal coming'. In India, new renewable energy does not necessarily displace fossil fuels because overall demand is expanding so rapidly. As the staff member at WRI explained to us, renewable energy 'adds low cost power to their portfolio but so far I have not seen it result in retirements of older assets because demand is also growing so its both a fossil fuel and renewables kind of track which is going on'. Indeed, to support the ongoing viability of coal, this staff member told us that, like in the Australian market, renewable energy generation is subject to curtailing, despite officially having 'must run' priority in the market. This is to maintain the capacity utilisation of existing coal, ostensibly for grid stability reasons; however, it also underwrites their thermal revenues.

Institutional design on the generation side is geared to delivering the vast bulk of cost savings from large-scale renewables to the distribution companies: the use of 'reverse auctions' for 'price discovery' in PPAs sees private sector generators compete on price to be awarded long-term contracts with the distribution companies. The success of reverse auctions in driving down costs has, however, come under criticism from other actors for favouring larger over small developers. This is because bids have minimum capacity requirements that in effect exclude smaller installations from bidding for contracts. This particularly impacts wind energy given the more decentralised model of wind energy in Karnataka. In an interview, a representative of the Indian Wind Power Association stated that 'What the governments do, they want to create a headline, so wind has become the cheapest in India – they don't understand by bringing down the tariff … they are not seeing the social impacts…'. The representative argued that 'the small players will never be competitive. You need to be at least 250 or 300 MW in size requiring high investment so who can do this: Goldman Sachs, Softbank etc.'. In sum, 'reverse auctions and giving everything away on a platter to the big guys'. However, from the perspective of SECI, concentration of ownership is not a problem in itself, provided that bids for PPAs were well subscribed and sufficient for efficient price discovery. 'Large scale concentration of ownership' in which there were 'maybe 20 players there in the market for the entire Indian market' was described as 'OK' because 'we are getting 10 or 11 bids for the large [PPAs]'.

Our discussions with developers revealed a range of strategies in a highly uncertain context. In addition to capturing opportunities for profit, some are driven by a desire to diversify portfolios across technology and geographical markets, including as 'loss leaders'. They may be more interested in securing a long-term (twenty-five year in the case of the Pavagada PPAs) 'hedge' against the decline in fossil fuels, and against the emergence of more advanced renewable energy technologies, than in securing margin. The desire for a renewable 'safe haven' for investors puts a premium on PPAs and also on collaboration with public authorities that can offer long-term stability and confidence against the vagaries of technological change and market volatility. The drive for 'reputational capital' gained from participating in the process of decarbonising power combines with a desire to acquire a profile and foothold in the emerging energy system. With this, private corporate renewable-generating outfits may be more than willing to bid to supply distribution companies below their minimum rate of return on capital.

Distribution companies reap the benefit of this along with the state renewables agencies such as KREDL that then charge a processing fee to developers as well as an ongoing access to infrastructure and leasing of land. Benefits for local landholders, who secure income from land purchase and leasing, depend on their bargaining power, which as we will demonstrate is often minimal. As we explore through our study of Shimoga wind and Pavagada solar installations, at the 'bottom' of the supply chain are those that lose the most from renewable energy development – the rural landless who may lose agricultural livelihood and access to pasture lands. It is instructive that value flows 'upwards' in a hierarchy from dispossessed landless to land holders, to state agencies, to private developers, and to monopsony distribution companies. However, this distributional dynamic is also challenged as local landowners and landless people, as well as state agencies such as forest departments, become aware of the substantial surpluses being reaped and seek to capture a larger slice of the pie.

At the installation level, land ownership becomes a deciding factor though as highlighted there is a profound difference between the impact of solar power and wind power. While solar may allow mixed land use with arable or pastoral use coexisting with PV, in practice solar panels often require exclusive land access and come with high-security fencing that literally encloses the land and protects the panels against potential incursion. Wind farms, by contrast, can be more easily made compatible with mixed use though they may cause other forms of disruption. Against this broad regional renewable energy institutional governance backdrop, we analyse two major cases and a significant proposed development to draw out some of these on-the-ground issues.

Shimoga Wind Power

Renewable energy in Karnataka initially centred on hydroelectric power and then, from the 1990s on small-scale wind power. Significant solar power came much later. Wind power began in 1996 with small-scale power mainly for factories. The Karnataka state government and the Ministry of New and Renewable Energy (MNRE) introduced household and business feed-in tariffs in 1993, which encouraged land holders to establish wind turbines for the grid (Panse and Kathuria 2016: 426). Turbine owners would pay a small Panchayat tax to cover basic amenities (unique to Karnataka) as well as the standard 2% corporate social responsibility (CSR) payments to the state government. A thriving local wind industry emerged based on small-scale India-made turbines. By the end of 2007, there were 438 wind farms with an installed capacity of 941 MW, with an average turbine cluster size of 3 (KREDL 2023). Most of the turbines were providing around-the-clock power mainly for commercial use, in textile factories, for instance. Investment was supported by tax concessions from the state government that allowed substantial upfront depreciation that delivered large tax write-offs. One energy expert described some of these early investments as 'all for tax'. As in other jurisdictions internationally in the same period, these wind farms, many of which were small and non-operational for long periods, could be better described as 'tax farms' (Knuth 2023).

In response to the 2008 federal renewable energy initiative, the Karnataka renewable programme from 2010 shifted the focus to larger-scale wind developments located on state-owned forest land mainly on uncultivated high-wind ridges. These larger corporate-run wind farms were often established by private developers who would secure land access, build the turbines and transmission lines, and operate the infrastructure with a view to on-selling them as assets to willing investors while retaining an operational and maintenance role. As such, these corporate wind farms created a new asset class for capital accumulation. On the ground the maintenance costs could be minimal, with returns only constrained by the conditions of the PPA governing the unit prices of power supplied.

The wind farm investigated for this study, the Kudrekonda wind farm, conforms to this model. The wind farm was started in 2007 and completed in several stages through 2011. It was built by the Indian power company Suzlon and encompasses 50 wind turbines, each about 150 metres tall, along a forested ridge line known formally as the 'Honnāli Forest range' to the North of the town of Shimoga in Karnataka.

There are four villages in the region that are most directly adjacent to the wind farm: Belaguthi, Chinnikatte, Surahonne, and Sovalanga. All are rural areas in the Davanagere district and Honnāli sub-district. The total population in these villages is 14,718. Honnāli is classified by the Karnataka government as a 'backward' *taluk*, meaning it is considered more developed than Pavagada, which is classified as 'more backward' (with 'most backward' the least developed), but behind other, predominately urban areas that are classified as 'developed' (Annigeri and Hanagodumath 2018: 12). This classification of the *taluk* is reflected in Census indicators for the four villages, such as literacy, the prevalence of agricultural substances, and the proportion of scheduled caste and scheduled tribe communities. On each count, affected Shimoga villages register less disadvantage than the Pavagada villages. The literacy rate in the four villages is 70%, which is lower than the Karnataka state average of 75%, but just above the average rural rate of 68%, and well above the literacy rate in the Pavagada villages of 55%. Twenty per cent of the population are from scheduled castes and scheduled tribes – 14% and 6%, respectively – compared with 41% in the Pavagada villages (Government of India 2011).

Of the 14,718 people in the four Shimoga case villages, 1,990 were landowners and 1,929 were landless labourers, representing one-third of the working population, respectively. The other third of the working population are in non-agricultural occupations. This is a much higher proportion than the Pavagada villages, which are more dependent on land for livelihood, with lower levels of non-agricultural employment. As with Pavagada, the majority (85%) of landowners are men, but unlike Pavagada, where women are the majority landless labourers, a smaller majority of landless agricultural labourers (65%) are also men (Government of India 2011). In contrast with other parts of Karnataka, the region is relatively fertile and landholdings are relatively evenly distributed. A long-standing environmental NGO representative, Ajit, highlighted the contrast in history between the Shimoga area where the land reform movement had been very active in the post-independence period and the Tumkur region where the Pavagada solar farm is located, as discussed later, where land holdings remain very stratified.

The area around Shimoga is dominated by farming and forests; it borders the Western Ghats mountain range and the Shettihalli Wildlife Sanctuary. Areca nut plantations and paddy fields are prominent all around the city of Shimoga, and forest ranges rise comparatively abruptly out of the plains, with a height of 100–200 metres above the plains. The topography ranges from tropical forests to grasslands and contains rich biodiversity value. The forests were considered sacred when their management was community based (the name Shimoga is very likely based on the words 'Shiva mukha' or Shiva's face/mouth). They were later integrated into forest resource planning as control shifted to the Forest Department (Ramachandra et al. 2013). A walk in the forests near the windmills illuminates the descriptions of the villagers we spoke to. The soil underfoot is of a rich red hue and rocky; the vegetation is diverse, ranging through verdant shades from evergreen to deciduous. There are beautiful small bodies of water, sometimes marred by the inevitable plastic wrap or bottles brought by trekkers and day trippers. The air is redolent with the fresh scent of greenery and the sporadic calls of birds. Even the windmills, visible through the mists on the hilltops, take on a mysterious and fabulous quality as if these artefacts of technology can only add to the sacred aura of the landscape.

Rakesh and Mallikarvind (all names are pseudonyms) do organic farming here, with about a dozen windmills within reach. Rakesh and his family of eight farm 4 acres of land, Mallikarvind and his family a similar size. They have the knowledge of six generations and thousand years of farming. The wind turbines are located at the ridge and clearly visible. Rakesh and Mallikarvind, however, have other problems than the wind turbines. There have been dry years and very wet years, resulting in little crop or washing away seeds. The wind turbines, they reckon, have little to do with this. Wind turbines are more or less an issue for the use of the forests, they tell. Since the installation of the wind turbines, the forest is regularly patrolled by the company Suzlon. For Rakesh and Mallikarvind, however, this is not really relevant, since they have not been going into the forest much. There certainly are shrines and the forest and some people may go with their animals to the fringes for grazing or collecting wood, they say, but there are also wild animals and not much reason for them to go to the forest regardless.

The company that patrols the forest and sets up not only the wind turbines but a substation at the foot of the hills, Suzlon, had been established in the 1990s. Suzlon grew to be a major wind company, manufacturing, installing, and operating turbines: in 2023, it was responsible for more than 100 wind farms with 14 GW capacity in India (Suzlon 2023). Suzlon's corporate communications frame renewable energy, as so many do in the sector, in a spatially unfixed and temporally flexible manner. In an interview at Suzlon's headquarters in Pune, Maharashtra, an engineer stated, 'Wind is the future. It doesn't need anything except the mills and doesn't shut down at night'. However, in practice, those windmills require the appropriation of land, which, in the case of Shimoga, was forest land.

The wind turbines at Kudrekonda were built on land leased from the Forest Department. The Forest Department official informed us that their policy was to favour leases for wind turbines, given they are not significantly disruptive, though the cost of leasing land had been increased from a relatively negligible amount in the early period. The forest is designated as reserved forest, requiring federal approvals for clearances. These had been obtained from the national-level Forest Department via Karnataka counterparts. A total of 199 acres was leased in various parcels for access roads and turbine foundations. The first, in 2007, was for 30 years and stipulated a 30,000 Rs. lump sum per turbine with offset planting elsewhere in the region. There were also direct negotiations with local landholders to buy land for the power infrastructure, including for a substation and transmission lines. There was controversy as the price offered was initially well below local land value, though subsequently, following local pressure, this was increased. A rough access road to the top of the ridge was constructed as well as service paths between the turbines.

Rates of return for wind power investment at Kudrekonda are very clearly spelt out in an application to the United Nations' Clean Development Mechanism (CDM) by Sargham Retails, an owner of 10 Suzlon-supplied turbines on the ridgeline, for approval to generate carbon credits from the project. As part of the CDM process, Sargham detailed project financing and the 'criticality of CDM funds for the viability of the project' (UNFCCC 2009: 16). Energy from the turbines was to be sold at Rs 3.40 per KWh over 10 years (and then Rs 4.08 per KWh for a further 10 years) to BESCOM, Bangalore Electricity Supply Company Limited. According to Sargham, this would deliver a rate of return of 9.22%, below the Bank of India lending rate at the time of 12.25% (UNFCCC 2009).

Certified Emissions Reduction (CER) certificates earnt from the CDM were to close the gap. At the time CER prices stood at about US$10, rising to $15 in 2010, crashing to nearly zero until recovering to US$5 by 2020 (Michaelowa et al. 2021). Sargham's turbines at Kudrekonda had an installed capacity of 13.75 MW and the company claimed these turbines would generate 26,475 CO_2-e in annual emissions reductions. This would net US$264,750 annually at the then US$10 CER price. Despite the collapse in CER prices, the company applied to renew the registration in 2017 and at that time estimated an increase in avoided emissions to 27,955 CO_2-e per annum (UNFCCC 2017a). Remarkably for our comparison, two German energy companies, alongside other coal- and gas-fired power plants and oil refineries across Europe, surrendered CERs from the Kudrekonda wind farm to offset emissions: In 2009, Vattenfall's Reuter West coal-fired power plant in Berlin surrendered 900 CERs from the project to comply with its obligations under the European Union's Emissions Trading System, and in 2017, energy supply company ExtraEnergie voluntarily cancelled 49,350 CERs from the project to provide 'Climate-neutral natural gas products' (UNFCCC 2017b; European Commission 2013).

Sargham's application also commented on local impacts, describing employment of local people, lack of displacement, and minimal other impacts.

With regard to land impacts, it commented that 'The land on which the project activity takes place is largely unproductive. Prior to the project activity, most of the land had no beneficial use' (UNFCCC 2017a). Further, it suggested 'the project activity has received complete support from the local populace…'. In order to comply with the CDM process, a stakeholder meeting was held in 2008 with several local representatives of Suzlon with two members of local Panchayats. At the meeting there was no 'negative or discouraging feedback from the stakeholders concerned'; the company added that 'In view of various direct and indirect benefits (social, economic, and environmental), all the stakeholders have supported the project activity' (UNFCCC 2017a: 38).

The company and its contractors were not required to gain approval from the elected *Panchayats* that represented villages on the lowland surrounding the wind farm: it was a 'higher-level deal' as one villager put it. In addition to the most directly affected villages in the Honnāli *Taluk*, the northern end of the wind farm is located in the neighbouring Shikaripur *Taluk* in the Shimoga District. Other than Ramesh and Mallikarvind, who saw little conflicts with the wind turbines, one panchayat member in Shikaripur *Taluk* reported that local people had objected to the project proposal but had been told the company would 'go ahead with installation, whether you like it or not'. The company had installed a pilot windmill, and when the full project was approved, a 100-strong delegation to Suzlon was organised with representatives from every household, raising concerns. In response the company had made a one-off donation to the local temple (1 lakh (100,000 Rs.)). Resentment over the wind farm and its impact on the forest remained.

In contrast to the developer's insistence of the lack of alternative use values, several local people stressed how the forest ecology and local culture are interwoven. One leading member of the local Panchayat, Arun, stated the 'forest mountain belongs to all these villages'. Another, Sanjay, emphasised the forest and its common lands as the foundation for village life: 'If there is no common land there is no village'. There are shrines on the top of the ridgeline dedicated to local deities that protect nature and the forest. For several the forest held strong cultural significance: when asked if they felt connected to the forest, Mahesh, a villager, responded, 'Yes, we are. Nature is our life. Our God is there on the hill'. The relationship is lived in other ways: one farmer, Sunil, stated that, philosophically, 'nature does the farming, not farmers, not scientists'. From this perspective, for instance, local 'Desi' cows are not simply cattle but are part of culture, with local Amritmahal and Hallikar breeds valued for the quality of their milk and dung. Goats may also have significance: in one village a goat with orange-painted horns was being readied for sacrifice at a Diwali feast. With this, 'nature' and culture can be correlated in ways not allowed for in formal regulatory processes.

When the wind project was first proposed, the four local Panchayats were proactive, offering their support if the government provided funding for youth training and enabled access to electricity from the project for the locality. They were told to wait:

> They said after 20 years, only then we can think of it.... They have taken forest common land for producing energy. We demand that the farmers get ... access to current throughout the day.... We feel that farmers were ... kind of cheated. We did not get any profit. There is environment, which is negatively affected. Forest area which was pristine, is now turned into kind of traffic area. (Interview Jayant November 2018)

When asked 'Did the villagers allow the windmill in the hope that they will get power one day?', the response from Anirudh was 'Yes, we believed them. We are farmers, and trust is in our nature'. Local people were disappointed, but more than a decade later were still hopeful:

> We believe that one or the other day, we might get electricity benefit. Revenue and forest ministers have made an agreement with the company. It is good that they are generating more power and build roads across the forest. But they also need to think of the welfare the citizens. (Interview Anirudh November 2018)

The issue had been magnified by visible CSR spending at Tuppadahalli, a nearby wind farm wholly owned by the Spanish multinational Acciona. This was constructed in 2012, and the negotiating position of locals was stronger. As farmer Girish put it, suggesting a process of learning through struggle:

> They knew what to ask for. Here nobody knew, they didn't know what to ask for. They had a better negotiation. This was the first in Karnataka on this scale. We did it even know who to negotiate with, company or government. This was the first in this area. (Interview Girish November 2018)

Unlike for solar power, there are very few landholders who have sold land or have been receiving rental income. It may be argued, by the same token, that there is no real cost, as reflected in the general disinterest in the project. But underlying the disinterest is an awareness that others are profiting. The windmills are a reminder of this appropriation and irrelevance of the local community, many of whom see the mountain as their cultural heritage, that rightly 'belongs' to the surrounding villages. In a theme that is repeated in Pavagada and indeed our German case study, villager Sudhakar stated, 'the power is generated here and being sold to big cities and to other states; this project is not useful for any of the villagers ... it is of no use to the people'. A woman, Neela, who had been a Panchayat member summed up the context:

> Around 10–15 yrs ago windmill was installed. They did for themselves. But for villagers it means nothing, no benefit.... If we work, we get money. If not, there will be nothing left to eat. Government doesn't care about us really. We have old-aged people, handicapped,

> single women in our village. No school, no health facility. People in power would say they would help. But they do nothing. We send our children to private school … we have to pay per child 1000 rs. for transportation and pay 6000rs. school fees. The government is not doing anything adequate to improve the living conditions of the citizens. Our children's parents work as daily wage labourers, their children also are forced to clean cowdung, be labourers in the future. There are no laws and no enforcement to really protect us. (Interview Neela November 2018)

She outlined how women work as labourers six days a week at 200 rupees a day whenever the work is available to pay for basic necessities. She was asked if electricity for irrigation from the windmill would help and responded: 'They have not given to us. They will never give it to us. There is no "if". Ours is a small village. They will never give power to us'. This question of power for villages and farmers was critical. The sense of cultural ownership translates into a sense of social marginalisation and indignation, as Ramesh, a Panchayat member, put it: 'They do not give electricity generated from the windmill to our houses nor fields. The hill is ours. This is our biggest grievance'. Hari, an older Panchayat member, stated: 'It is unjust … villages and farming land should have got access to power', and argued this could even be a model for windmills: 'if we get power especially for the farmers, they can replicate this everywhere'.

The forest, in formal terms, is state-owned and provides the ideal 'unused' landscape to be allocated for wind power. But all land is socially and ecologically 'used' in one way or another, and no less so for the Honnāli forest, which has its own ecology and, as noted, its social, cultural, and spiritual uses by surrounding villagers. Certainly, social use of the forest had also changed with the wind farm, beyond simply electricity generation. The top of the access road to the turbines is at 820 metres above the surrounding landscape and is spectacular, especially at sunset. Google Maps describe it as 'windmill wonderland'. It attracts devotees and tourists who load up their Instagram photos and TikTok videos; the turbine columns have graffiti including declarations of undying love. Internet reviewers describe a 'very silent place' with a 'very peaceful temple'. The access roads and connecting pathways have effectively opened up the forest to wind turbine tourism and social gatherings. As Ravi, an older Panchayat member, stated 'It was a pristine forest. Now there are many people going around for picnic'. Another, Ganesh, gave a less charitable assessment, raising concerns that people had 'started using [it] for drinking, plastic waste thrown, gambling, lovers misusing the place, destroying the nature'.

At the same time, surveillance had increased. The company had employed local people to work as security guards, to protect the infrastructure. These guards, four or five at any one time, would patrol the site. Local people were aware of this and some were discouraged from entering the forest, which had become noticeably more dense. The access roads had changed how the forest could be used.

According to Murugan, from a nearby village, 'now the road is made, and the patrolling jeep moves around, so the freedom to roam around in the forest is curtailed'. The Forest Department confirmed that increased security had discouraged organised wood collection (as against household foraging or other forms of forest use), which they stated had been commonplace. In part this was because local people no longer needed to burn wood for cooking, due to the introduction of gas cylinders (as part of a national scheme). Local people stated they had less need for informal access, to gather dead wood for instance; some stated they still accessed various forest products and used them for grazing, with no objection from the Department (though formal access was not available). Several said there had been an increase in wildlife, including tigers and cheetahs, which threatened stock. Despite this, the return of denser forest and wildlife was generally welcomed by local people, as signs of a healthy thriving ecology.

We visited Venkatesh, a larger landowner to the west, who had 100 acres of betel nut and coconut under the wind turbines. He only mentioned the noise of the turbines, commenting it was generally drowned out by everyday village noise. We had several conversations with small-scale farmers to the East, including organic farmers who are part of wider state-wide networks. Here landholdings were very rarely larger than four acres, and generally one or two acres per household, and freehold. Those with smallholdings rely on agricultural labouring. Rain is regular and the land is fertile, sustaining two crops a year and a range of grains and vegetables; households also have livestock, invariably goats and local breeds of cows highly prized for their milk and dung and for their cultural significance. The local village is highly diverse, with a wide range of castes, Kurubas (traditional herders), and Lingayats (high castes who reject caste) described as dominant. Landlessness was commensurably low, with landless pastoralists visiting the region at different times of the year.

One village we visited was made up of the formerly nomadic Banjara community, a Scheduled Caste in Karnataka. The community of 80 households had been there for four generations following government schemes encouraging them to settle. About four-fifths of the villagers have access to one to two acres of land near the forest, with permission to farm without ownership, called 'Bagair Hukum' in Karnataka; the rest had freehold title, a few with larger plots, up to 15 acres. There is an effort, via the local forest committee, to convert the 'Bagair Hukum' lands to freehold, requiring an extended process of community negotiation with the forest authorities. A villager here, Surya, contrasted the legal hurdles they faced as a community with the wind power company, which had access to the Chief Minister and accessed forest lands with relative ease. However, they also stated that the road paved to the top of the ridge, serving the construction and maintenance of the wind turbines, is also of advantage for locals visiting the shrine located close-by on the mountain range. It can now be accessed by car as well as by foot, making it easier to reach for less mobile devotees.

Yet, easy and fast transport also causes suspicion: Local villagers observed outsiders and helicopters that had recently been undertaking prospecting surveys, later revealed to be prospecting for gold. Certainly, the possibility of a goldmine in the region, with allied problems of groundwater and arsenic tailings, poses a much greater threat to the local communities and demonstrates the extent to which the advent of windmills was hardly the most concerning problem facing the locality.

Across our ethnographic fieldwork, there was a striking absence of any sense that wind power was part of a solution to climate change. Many understood climate issues but believed that wind power would be additional to existing coal-fired power, not at the expense of it. Several farmers were highly knowledgeable about the changing weather patterns, which were already having an impact on livelihood, notably with the 2019 Karnataka floods, but also more generally with unseasonable rains. This posed a major problem for farmer Balu, who was planting fruiting shrubs and nut trees more able to withstand the weather. There is extensive debate about how best to respond in what is a rapidly changing context. Locals generally linked this to global warming but did not assume renewable energy was addressing this. Certainly, the wind farm had never been presented to them in this way.

Likewise, and in stark contrast to the politics of wind power in Australia and Germany, no one raised any concern about the impact on the aesthetics of the landscape. One farmer, Vijay, said, laughing, ‘to look at it once, it is nice … now it is normalised’. While talking with parents and primary-age children at a temple, we asked whether they liked the windmills and a child said: ‘We are happy because it is in our village. We don’t know if it is good or bad for us’. They had learnt about it in school lessons: ‘We know we shouldn’t touch electricity’.

The Pavagada Solar Park

The Pavagada solar park is in the Pavagada Taluk in the district of Tumkur, about four hours’ drive north of the capital of Karnataka, Bengaluru. It is one of the largest solar projects in the world, occupying 12,000 acres of land. The solar farm encompasses five main villages (Balasamudra, Thirumani, Kyataganacharlu, Vallur, and Rayacharlu), centring on Thirumani, and supplies 2,050 MW of installed capacity to India’s national grid. It was established in 2015 through a collaboration between SECI and the state government-owned KREDL. As already outlined, the two agencies established a joint venture company, the Karnataka Solar Power Development Corporation Ltd. (KSPDCL), to manage the development of the project.

Pavagada was chosen for the solar park because of its hot, semi-arid climate on an elevated plateau, its drought-affected agricultural economy, and the relative poverty of the population (Jain 2015). Pavagada has minimal social and health infrastructure, poor schooling, a low literacy rate, and one of the lowest levels of college attendance in the state. Pavagada is therefore considered a 'more backward' *taluk* by the state government's 'development index' (Annigeri and Hanagodumath 2018). The literacy rate is 55%. There were 10,294 people living in the five villages that surround the solar park at the time of the 2011 Census. There were 2,257 cultivators (i.e. landowners) and 2,580 were agricultural labourers (i.e. landless workers) (Government of India 2011). This is a higher proportion of land dependence for livelihood compared to the Shimoga case. Whereas in the Shimoga case villages 32% of the working population were in non-agricultural work, in Pavagada it is only 12%. Traditionally, the landless populace works as agricultural labourers for landowners and grazes stock on private and common lands. Most landowners had small-to-medium holdings. About 155 landowners owned more than 10 Ha (about 25 acres); 1,550 owned between 10 and 1 Ha (about 2.5 acres); and 900 families owned less than 1 Ha (Knight Frank 2016: 38).

The stratification in land ownership intersects with caste and gender inequalities. While most landowning farmers are men (1,523 men; 734 women), the majority of landless agricultural labourers are women (1,409 women; 1,171 men). Around 41% of the population in the five villages are from Scheduled Caste or Scheduled Tribe communities (Government of India 2011). Pavagada has been the site of conflict over land redistribution. This reflects local inequalities and relative disadvantage, and also its location near the border with Andhra Pradesh where there has been a long-running Maoist insurgency.

Agriculture in Pavagada has become increasingly precarious due to limited rain, leading to indebtedness, migration, and suicide among farmers. It has been officially declared drought-hit 54 times in the last 60 years, and the Central Water Board has declared its groundwater is 'overexploited' (Prabhu 2017; Rajendran 2019). Several officials we interviewed in Pavagada described the region to us as a 'virtual desert' or a 'wasteland' in making the case for the siting of the solar park. During our fieldwork in 2018 and 2019, we observed wide variation in agricultural conditions.

One can see windmills on the hills in the distance from the Pavagada solar park, faint shadows shimmering in the heat haze. Prior to our last visit in late 2019, there had been substantial rains, and the landscape looked relatively lush. Farmers pointed out that this appearance belied general water scarcity and the rising impacts of climate change. However, despite the official discourse on failing agriculture due to lack of rains, there are pockets of verdant greenery in the region. We visited, for example, a well-established organic mango farm in a plot surrounded by solar panels. It was replete with profuse fruit trees and lush vegetable plots. Puneet, the owner of the farm, had relatives who had leased land for the solar park, but he was invested in being a farmer. He pointed out that the aridity was partially due to the fact that levels of irrigation were lower than comparable arid regions. One of our interlocutors, who had leased a substantial plot of land to the solar producers, took us to another patch of land owned by his brother. The brother, Paresh, had refused to lease and continued to cultivate his plots right in the middle of the solar panels. These fertile green holdouts against the lure of leasehold money complicate the general rhetoric of the land being poor and unproductive and show that these official descriptors are contested locally and influenced by outside political and economic interests.

Decisions by farmers on whether or not to lease their land are indicative of the Pavagada solar park's innovative land leasing model. This model potentially represents a departure from other models of land acquisition used in India for industrial development. Indeed, there is increasing evidence that large-scale solar developments in India have involved compulsory or coercive land acquisition resulting in displacement of communities (Stock and Sovacool 2023, 2024). The issue of land was also complicated by the fact that Pavagada was near an area where there were previous Maoist insurgencies over the issue of land use and ownership and therefore the population was considered more 'politicised'.

Government officials in Karnataka, cognisant of conflicts over 'land grabbing', proposed the land leasing model as a means of voluntarily enrolling local farmers as long-term 'partners' in the project. In doing so, they were strategically responding to the distributed and uneven land ownership patterns in the area. Land aggregation (or 'banking', as it was described to us) was necessary because no single landowner could provide the quantity of land needed for a large-scale solar park. Leasing became the solution for this land aggregation problem because many landowners were unwilling to sell, due to long-held connections with the land. As a large landowner, Anmol, who led negotiations, explained to us:

> Whatever maybe the extreme condition, we are never willing to sell. It is kind of an attitude problem. It is an asset. If my father had sold, I wouldn't be having anything for me. It is your responsibility to maintain it. (Interview Anmol November 2019)

At the opening ceremony for the park in March 2018, the Minister emphasised that the project treated farmers 'as the key partners, as also beneficiaries', and was 'creating new job opportunities and economic growth leading to the prosperity of the people of Pavagada' (Energy World 2018).

The land leasing model was established as an entirely voluntary opt-in arrangement for local farmers who would retain ownership of their lands. Leases of land for the solar park were successfully negotiated within six months of landholders first being approached. Most landowners in the five villages leased their land to the project – 1,422 in total (Knight Frank 2016: 38). The standard leasehold is a 28-year lease agreement that sets a base rate of Rs. 21,000 per acre per year (about US$300). This rate is indexed at a 'simple' interest rate of 5% every two years, with indexation based on the original principal. As such, interest amounts to an additional payment of Rs. 1,050 per acre every two years for the term of the lease. After 28 years, the lease can be renewed under mutual agreement between the landowner and KSPDCL. Otherwise, the land is to be returned in its original condition to the landowners, and there is a commitment to rehabilitate the lands for agricultural use, including to remove the solar panels and their concrete footings.

The leased land was levelled, and all vegetation removed before being divided into plots that could host 250 MW of solar generation capacity each. The lease provides exclusive access to the land for KSPDCL and for the developers who sub-lease it. Developers enter a 25-year sub-leasing arrangement with the KSPDCL, with periodic land rent, operational charges, and development fund contributions, creating a pool of funds to provide for infrastructure and local development needs (KSPDCL 2015 & 2020). KSPDCL, as broker for the land leases with both farmers and developers, has therefore been able to establish a new very large plant at very low public outlay.

Once access to land was secured via the land leasing model, plots were allocated to private investors through a competitive bidding process for PPAs. As we have discussed, this is an increasingly common policy model globally that seeks to both drive down the price of power and provide a government-backed income stream that developers can use to access capital (Bridge et al. 2020; Baker 2022). Twenty-five-year PPAs to supply electricity at a fixed rate per unit were 'reverse' auctioned to eight Indian and foreign large multinational companies by national and state agencies. The initial investors included Tata, Fortum, Adani, ReNew, Avaada, Softbank, and ACME, as well as KREDL, which maintains 50 MW of capacity.

The price of PPAs auctioned off dramatically declined in a short period of time. The first set of auctions in 2017 resulted in contracts being awarded at 4.8 Rs/KWh. The cost of the power supplied to the grid in reverse auctions conducted in 2018 subsequently fell to 2.8 Rs/KWh. The contract for the final 50 MW operated by the state-owned power company KREDL, negotiated in 2019, is even lower, at 1.2 Rs/KWh (KSPDCL no date; KREDL 2019). These prices are lower than average power purchase costs for both fossil fuel and other renewables paid by Karnataka's DISCOMs. The cheap power supplied by Pavagada is significant due to the high levels of financial stress and indebtedness of the DISCOMs. Lowering the purchase cost of power has made the subsidies provided to unmetered rural electricity users more financially sustainable for DISCOMs. This reflects a strategy identified by Kale (2014: 9) in states facing strong rural demands for electricity subsidies, which resist market reforms by 'finding alternative ways to augment electricity supply and utility profitability'.

High investor interest despite rapidly declining prices shows the Pavagada model is an attractive one for private investors, who are anticipating a growing renewable energy market and falling technology costs. The model offers a long-term planning horizon with fixed costs and income and low levels of forward pricing risk. The guaranteed power purchase price for 25 years is a major incentive in an international context where prices in PPAs are often varied over time. Investors were also attracted to a new, innovative land leasing model that was promoted by KSPDCL as an opportunity to participate in what could become a more widely used model in the Indian solar economy.

Like Shimoga, developers at Pavagada sought to augment their revenue by claiming and selling carbon offset credits for the emissions reductions produced by the project. Finnish state-owned energy company Fortum registered their 250 MW plot at Pavagada for carbon credits, initially with the UN's Clean Development Mechanism, and then with the Gold Standard voluntary offset standard programme. The documents establishing the project contain commercial information not otherwise available (UNFCCC 2020). They calculate that, prior to offsetting revenue, the post-tax internal rate of return on equity would be 6%. However, developers of offsetting projects have an incentive to claim lower rates of returns in such documents to prove the need for offsetting revenue to achieve 'additionality', and during our field work an estimate of 10% return was commonly quoted by government officials. Indeed, as renewable energy has become economically competitive with fossil fuels, solar and wind projects, especially those developed by large energy corporations, have come under increasing scrutiny for their additionality claims (Hodgson 2021). By the time the project was registered in 2020, demand for carbon credits had shifted from compliance to voluntary offsetting markets, driven by restrictions on the use of offsets in European and other emissions trading schemes and the rise of voluntary corporate Net Zero pledges. Offsets from Pavagada were voluntarily surrendered by gas- and coal-fired power plants owned by Vattenfall in the Netherlands, as well as the IMF, the European professional golf tour, and a logistics warehouse in the UK, among other anonymous users (Gold Standard 2024).

The Pavagada Experience

On the face of it, the Pavagada solar park contains the potential for significant social, economic, and environmental benefits for landholders, the Karnataka government, investors, distributors, and the climate. Reflecting this, the Pavagada model has been presented as an exemplar for other Indian states to follow. Does the Pavagada model live up to these high expectations? This section examines the local community's experience of the solar park, drawing on our ethnographic evidence from interviews, workshops and meetings, and other relevant documents. We analyse the socio-ecological relations and outcomes of the project for landowners, landless labourers, employment, women, and the local environment, focusing on the views expressed by local people during our fieldwork.

The need to take proactive steps to ensure positive outcomes for the local community was recognised by social and environmental impact documents produced by the World Bank. One report clearly identified that the landless population was vulnerable to the loss of agricultural employment and access to livelihood:

> One of the critical issues would be the impact of the project on the agriculture labour/landless and especially on the marginalised sections of the society including the [Scheduled Caste] and [Scheduled Tribe] population. (Knight Frank 2016: 40)

With the solar park now fully operational, our research examined the distribution of outcomes among different social and economic groups – identified by the World Bank prior to project development – in practice.

Outcomes have differed for larger and smaller landowners. The minority of landowners with larger holdings have, in general, leased the bulk of their land to the solar park and retained some land, usually irrigated, for agriculture. Leasing land for the solar park offered an opportunity for larger landowners to diversify their income. Smaller landowners were more likely to lease all their land and could not pursue the same diversification strategy. Previously, most farmers would earn between Rs. 15,000 and Rs. 25,000 per year per acre depending on the rains. Initially, the assured yearly lease rents of Rs. 21,000 per acre plus interest were generally seen as adequate. Given increasing weather fluctuations, this income was viewed as more secure than farming. In contrast to Brandenburg, where leasing land to wind farms was much more financially lucrative than cropping, the primary benefit for lessors was this promise of income security.

Landowners also agreed that there was no coercion to persuade them to enter the leasing arrangement. However, they emphasised that at the time of the negotiations, many were in serious financial stress. Negotiations were conducted after an extended period of drought and several farmers recounted that they needed a more reliable source of income. One landowner, Naresh, who plays a prominent role on a committee of farmers who have leased their land, explained:

> While we were struggling due to no rains since 10 years, the political leaders came, held meeting and it became one sided decision for farmers. They told they give this much … They came, they asked, we were hungry, we gave it to them. (Interview Naresh November 2019)

While leasing the land provides landowners with a much more secure income than rain-contingent agriculture, many now held the view that the annual rent did not reflect adequate compensation for the contribution their land was making to the project. This view emerged when landowners compared and calculated the economic returns from the solar park with the diminishing value of the leases. The simple, rather than compounding, interest rate in the loan means that rental incomes may not keep pace with inflation over time. Naresh said people were questioning why farmers:

> gave away for such low price for a [multinational company]. They have made so much benefit … At present they are giving us [in lease rent] one day of their earning out of 365 days. (Interview Naresh November 2019)

Landowners emphasised an adequate rent amount was needed because they were making a long-term commitment that reduced their alternative options as landholders. The re-zoning of land from agricultural to industrial use also has some potential financial disadvantages for landowners, who lost access to concessional agricultural loans. When asked what advice they would give to other communities being approached to lease their land for solar, landowner Arjun said:

> First, they should ask for increasing the rate. Then, while writing the agreement, they should show it to someone with education, and get it examined as to what has been included in the agreement. Because, I don't know how to read, I don't know English, I can only sign, that's all. If we have to understand the agreement, then we have to show it to someone like you. (Interview Arjun November 2019)

This comment attests to the power imbalances that structured the negotiations, which left many landowners feeling as though they hadn't been given complete information.

Disconnection from the land generated some ontological fears among landowners, an example of identity conflicts in renewable energy. They thought that they might lose skills and knowledge over the 28-year lease term, making it difficult for the next generation to continue farming. Several farmers were very concerned that it was no longer possible to identify their lands in the levelled solar park. Landowner Shiva reflected this sentiment:

> We feel our land was somewhere here, we are not able to recognise our own land … I have worked in my land throughout my life; still I am struggling to keep my field memory alive. Then how will our children and family members remember. It is like delineating one's family from our roots, from our history is what bothers me the most. (Interview Shiva November 2019)

Last, the shift from farming to land leasing generated more immediate material change in reducing the capacity of farmers to grow food for their own needs. Vishnu, a landowner who also works at the solar park, explained this shift, saying 'last year we had milk, buttermilk and curd at our house. Now we are buying'. Landowner Bhanu described this in terms of the commodification of social relations, via increased market dependency, representing a key point of difference with installations in our South Australia and Brandenburg case studies where land is not usually used for subsistence purposes:

> If we grew lentils for a year, we would use it for household consumption up to two years. Peas, millet etc, we used to grow and whatever is left after our share for household consumption, we used to sell it in the market. Now for every small thing, we are dependent of the market. (Interview Bhanu November 2019)

Despite its inadequacy, the lease income offers benefits by protecting the larger landholders, and to a lesser extent, the smaller landholders, from the uncertainties of agriculture.

As anticipated by the World Bank reports, local people who have no land, or have very little, are the main losers from the model. Describing the situation before the development of the solar park, one landless labourer, Krishna, said:

> I used to get work in the crop/agriculture season. You can say roughly five, six months of work. Only in peak summer, for four months there won't be agriculture work. (Interview Krishna November 2019)

The landless population has historically relied on this seasonal agricultural labour, as well as very small-scale informal subsistence cultivation and access to common grazing lands. Referring to relatively open access given to landless labourers for grazing outside of peak growing seasons, Narayan, a landless man, said that this had been disrupted:

> Before solar it was very good for our sheep. After solar there are difficulties…. There are companies around us. There is no space for grazing. (Interview Narayan November 2019)

The loss of local agricultural labour, and access to grazing land, has undermined the primary sources of livelihood for landless labourers, who are disproportionally Dalits and Adivasis (Scheduled Castes and Scheduled Tribes). One landless woman, Radha, described the loss of employment as a condition of 'idleness' and said they were surviving on small loans from larger landowners who had leased their land. Another, Usha, said this was exacerbating inequality because:

> People who had more land become wealthier. We are where we are, landless people. (Interview Usha November 2019)

This view was also widely held by landowners, who expressed to us their concern for the future of local landless people. A landholder with 3 acres of land, Arul, said:

> People who do not have land, no benefit for them. Rather, they have lost their livelihood, as no one calls them for work. The landed people have given away their land for solar plant. A village means not just the landed agriculturist. There are all kinds of people, like people with livestocks, labourers, small farmers, people doing other jobs, etc. The government should support the villagers. But the situation is bad. The landless, poor people are forced to migrate and work outside. (Interview Arul November 2019)

Finally, one of the biggest frustrations among landless people was not only the loss of agricultural employment but also the perceived lack of infrastructural benefits from the solar park. Local villagers claimed to us that people 'like you' had come to the village and promised schools, hospitals, and water and power upgrades in addition to the solar park, but that these promises had not been kept.

Like Shimoga, electricity was a particular cause of frustration due to the stark contrast between the solar and transmission infrastructure surrounding the villages and the unequal and unreliable connections to the grid within the villages. The poorest households face problems of unpredictable load shedding and supply levels that are limited in many cases to being able to power a single lightbulb or appliance. This is another key point of difference for India compared with our other advanced economy case studies, where local complaints about access to electricity generated in their area are usually focused on the failure to pass through the price benefits of renewables, rather than the connection itself. One of our interviewees from a local village, Bhupesh, said:

> Initially when they came, they said they'll give free current (power) and filtered water. Those people told us before they even took the land … They also said they will build hospitals. Nothing is given. Lot of promises was made. (Interview Bhupesh November 2019)

The social development funds created with the establishment of the solar park were very slow to disburse funds. During our fieldwork, we found no evidence of solar-funded social projects, which accorded with sentiments in the community. However, following pressure from community members, a journalist with *The New Yorker*, Meera Subramanian, who visited after our fieldwork, reported that some teachers, school buildings, and roads had been funded out of the social development fund or directly by international operators. These initiatives, however, were variable in quality and impact and did not directly address concerns around inequality between landowners and the landless. As Subramanian reported, 'landed neighbours now earn lease income, but the jobs are gone' (Subramanian 2023: 4).

Employment in the park itself, however, has been a key issue of contention. At initial meetings between officials and landowners about the land leasing model for the solar plant, the energy minister stated that the development would help address unemployment faced by farming communities in Pavagada. However, during the meeting at KREDL's office building in Bengaluru, described earlier, when asked if there are many jobs in the renewable energy sector, the managing director admitted 'No, not really, only during construction'. And his colleague added that the construction period in the renewable energy sector is too short for significant local job creation; 'it will be over before training for locals has even started'. The practical experience has been closer to that described by KREDL's managing director.

The solar park has generated some additional employment, but this has been unevenly distributed both among local people, and between local people and migrant workers from other parts of India. The project was projected to bring at least 1,400 direct new solar jobs (400 skilled technicians and 1,000 semi-skilled workers) to the region plus indirect jobs in hotels, restaurants, and vehicle maintenance (PwC 2015: 97). Local people stated there had been fewer jobs than predicted, and that many of these were only temporary, because they were limited to the construction phase, or intermittent, in the case of cleaning and grass-cutting work.

Limited ongoing job openings were generally allocated to landowners and their families. This reflects KSPDCL's Implementation and Support Agreement for the project, which states that 'first priority' would be given for family members of landowners who had leased their land (KSPDCL 2018: 19). Landowning farmers said that access to these jobs was crucial for them to supplement their lease income. Shipa, a landowning farmer with professional working experience, said:

> We need work. Otherwise we can't survive in the lease amount. Or else they have to increase the lease amount. (Interview Shipa November 2019)

However, this policy excludes many of the landless workers, who are not benefiting from lease rents, from employment opportunities. The experience has been a source of disappointment and frustration for landless labourers, many of whom reported being promised employment opportunities. Radha said:

> Some people from solar park side, met us and promised that first preference will be given to landless people. Glass cleaning work will be given to you people.... Energy minister promised work for everybody, especially landless people. (Interview Radha November 2019)

Workers, particularly security guards, which is one of the more consistent jobs available, report quite different levels of pay and working conditions between different companies operating in the solar park, which has resulted in regular labour disputes. Some companies are widely understood to be better or worse employers, based on their hiring practices, payment of wages, and access to other benefits such as health plans. This variation is in part due to the fact that working conditions and pay are not regulated via the main components of the Pavagada model: land leases and power supply contracts.

There is also general resentment from both landowners and landless labour for being excluded from higher-paid, skilled jobs by subcontracting arrangements that use internal migrant labour from states such as Orissa, West Bengal, and Jharkhand. This experience has clear overlaps with the experience of local communities in South Australia, despite the very different labour market contexts. Giri, who leased their land, said:

> Now they bring, labourers from other states since these corporate project managers, have their own vendors who supply labour, resulting in commission for the managers. When youngsters ask for a job, they are often denied it. We are giving lands but not getting a job in return. We have skilled labour among us too, including diploma holders, [industrial training institute certificates]. (Interview Giri November 2019)

Training programmes that were included in the Implementation and Support Agreement as a way of overcoming local skill shortages had also not materialised (KSPDCL 2018: 19). The key aspiration of local villagers to whom we spoke, especially local women, was for better jobs not just within, but most importantly beyond, the solar park.

The loss of previous sources of income, whether agricultural income for landowners or agricultural work for labourers, has disproportionally impacted women and their financial autonomy. For landless women in particular, the loss of open access grazing meant they had lost an important source of autonomous income. Similarly, many women who were small-scale landowners were also responsible for rearing sheep or goats, which provided a source of livelihood that they were able to exert some control over. Leela, a 37-year-old woman with eight daughters, said that she had leased two acres of land to the solar park but had been told that grazing would continue to be possible. She finds it increasingly difficult to maintain her flocks because the pastures and commons are now enclosed within the boundaries of the solar park. She was forced to sell some of her animals and migrate to nearby regions for four months of the year to graze her remaining animals.

Women we interviewed described how these changes entrenched gender inequalities. This issue was put to us forcefully in a meeting at a village where women conveyed their dissatisfaction with the loss in income. Radha said that since losing their sources of income, 'we have to depend on our husbands' income and they control us more'. Parallel issues emerged for women from landowning families. Women we interviewed protested that the process of negotiating land leases had placed male landowners in a privileged decision-making position. Radha said:

> They have not discussed with their families. The men went to meeting in Thirumani village. It was all men meeting! It was not a discussion over there, just agreeing to whatever they offered. (Interview Radha November 2019)

Women have also been less likely to find sufficient work to supplement rental income. Jyoti, a 31-year-old woman, said that the women in the village were excluded from the solar project as there were few jobs for women. She also mentioned that the male migrant labour who occupied the jobs available was viewed as a potential security issue for women, which enabled their families to further restrict their movement in an already patriarchal milieu. Sneha, a 35-year-old woman who had leased 5 acres of land to the solar park, commented:

> Earlier women had agricultural work in the village and used to earn Rs 200 for sowing, weeding, harvesting and so on for 4 to 5 months a year and used to get paid once weekly on Wednesday. Every Wednesday there is a village market and they used to buy household stuff with the money and generally led an average life. Now, there is no work anywhere. (Interview Sneha November 2019)

Women were particularly concerned about no longer being able to pay school fees and extremely worried about their children's future. One local landless woman, Geeta, said:

> We do not want our children to lead a life like us. We want to give them better education, so that they can make a life for themselves.... Our aspiration is to educate our children and we cannot afford to send them to convent. (Interview Geeta November 2019)

This perspective speaks to a much wider scale of implications for labour of the solar park beyond the installation's boundaries. In Geeta's case, she is concerned about whether the solar park is part of a model of development that could support her aspirations for her children to access education and find decent work in the economy more generally. There are also direct implications of the solar park for labour in other parts of the solar supply chain. Although our field work was focused on experiences in and around the installation, the manufacturer of solar panels sourced in the Pavagada solar park has been alleged to have used forced labour of Uyghur peoples in Xinjiang, China (Yunker et al. 2023)

Finally, there are several environmental concerns and conflicts. There are concerns that meeting the park's water needs, including regular cleaning of the 400,000 panels, draws on drinking and irrigation water supplies at unsustainable levels. The Social Management Plan warned that 'the withdrawal of ground water through bore wells to meet the [operations and management] demand for proposed solar park would lead to adverse impacts on the already deteriorating ground water conditions' (Knight Frank 2016: 54).

These impacts need to be compared with the water demands of pre-existing agricultural activities, both in terms of the quantity of water used and the relative social and economic benefits of that water use. They also need to be compared with more sustainable ways of managing the water requirements of the solar park. Despite recommendations in these reports for water recycling and rainwater harvesting, the water consent agreement issued by the Karnataka State Pollution Control Board (2016) allows operators to withdraw water from bore wells without any requirement to mitigate the impact. KSPDCL representatives stated that some developers were using a water-less cleaning mechanism, but since its use was not mandatory and was relatively more complex to implement, not all companies had adopted it.

There are also concerns about the management of the local ecology surrounding the plant. Again, water is a key concern, because solar park construction has altered irrigation infrastructure. Farmer Krishna said:

> The water level is too low, because to develop solar park, the company people have demolished the check dams. The rainwater flows away, they are not able to store the water and because of this the ground level water is also depleting. (Interview Krishna November 2019)

These impacts have been unevenly distributed by caste. Dalit communities have been most heavily impacted because they are excluded from using water in the main village and therefore depend on water supplied to common tanks (Saldanha and Rao 2020).

All of the farmers to whom we spoke were concerned about the future health of their land and its condition after 28 years of being covered by solar panels that absorbed such high heat. These panels are attached to concrete pillars with bases deep in the ground. Though KSPDCL officials assured us that, after the leasehold ended, the land would be returned to the farmers 'in the same condition as it was before', many farmers were not so sure. They felt that the land would require massive recuperation after being covered by solar panels and pointed out that they would have little recourse if the recuperation failed or was not undertaken. They were worried about the impact of land clearing and levelling on future agricultural productivity and were calling for money to be spent on afforestation. They were also concerned about the disposal of solar panels and other materials at the end of their life cycle and/or the lease term. Landowner Ashwin, who leased 20 acres of land, viewed this environmental question through the lens of power imbalances:

> We are owners, but owners with no control over our land … we have no control over the situation so will not be able to negotiate a better deal even after 28 years … eventually the government and the companies will become powerful. (Interview Ashwin November 2019)

Assessing Pavagada

Overall, the Pavagada Solar Park has delivered much-needed renewable energy to India's grid, which is critical for addressing the climate challenge. The innovative land lease model offers potential benefits for the development of solar power by addressing land bottlenecks while avoiding the need for controversial, often forced, land acquisitions. The current leasing and power purchase model is delivering substantial benefits for some stakeholders. Solar developers are guaranteed a return at low risk on their investments over a long-time horizon. State-owned power distributers are accessing a major supply of clean electricity at very low cost, potentially supporting the continuation of power subsidies for rural users.

Large landowners are able to diversify their sources of income with rental payments underpinned by the state government.

Yet, the Pavagada model is unequally distributing the benefits of renewable energy while entrenching existing social inequalities and generating distributional conflicts. The majority of the landless population has experienced a loss of income from agricultural employment on land that has been leased, which has not been fully replaced by access to new jobs in the solar park or elsewhere. Women have suffered the greatest losses because they are over-represented as landless workers and have lost income from agricultural work that was a source of financial independence. Both landowners and landless have experienced a loss of connection to the land that was a key part of their collective identity, posing questions about future livelihood and vocation, both for themselves and for future generations. Broader social and economic development opportunities that could be underpinned by the solar park are yet to materialise.

The Pavagada model, combining land lease aggregation and competitive bidding mechanisms for PPAs, is reflective of a globally dominant mode of energy transition that is primarily focused on driving down costs and managing risks for investors and states. There are, however, some relatively simple changes to the Pavagada model that could improve social, economic, and environmental outcomes. These include ensuring that land leases are properly valued and indexed, and that PPAs require local employment of landless people, prioritising women, and those from lower caste and Adivasi backgrounds. From the perspective of private investors and state actors, these represent costs and risks that need to be absorbed. However, this is necessary if Pavagada and other large-scale renewable energy projects that adopt a similar model are to generate real and shared benefits for local communities.

The difficulties faced by people in Pavagada are, of course, structured by a much longer history of exploitation and colonisation that is common across rural India. This is recognised by local community members. Bhanu explained:

> We are not saying that we have 100% difficulties because of [the solar park]. Earlier the crop yield was very less and now we somehow lead our life. But we do not have better life. (Interview Bhanu November 2019)

However, the Pavagada solar park represents a missed opportunity for transition that would leverage renewable energy development to create better lives for disadvantaged communities.

Over the course of our research, it was clear that communities had thought deeply about what would be required to achieve more transformative change. The most common idea proposed was to move towards a more substantively democratic partnership model. Key components of this model would be a revenue-sharing agreement and greater scope for the co-management of land to allow greater co-existence of electricity production and agriculture and encourage better employment outcomes. According to Bhanu, this was informed by a sense of permanency of Pavagada's role in the energy transition:

> The energy production will continue, the structures cannot be dismantled. Therefore, they should give us partnership. (Interview Bhanu November 2019)

Conclusions

The KSPDCL official who hosted the research group at the Pavagada solar park in 2019 described it at the time as a 'university'. With the solar park, he explained, the Karnataka state was experimenting with how best to achieve a transition to renewable energy. The conception of the project, in a drought-prone region with multiple landholders, was seen as a new departure for renewable energy policy nationally, as well as for Karnataka state. There were multiple lessons to be learnt from it – in terms of engineering but also in terms of institutional innovation, land use and environmental change, and in terms of social engagement. The idea of a solar park as a 'university' in this respect, with multiple 'disciplines' brought to bear, is revealing of the dynamic in play. In conversation with the KSPDCL representatives, we asked whether they thought the land would continue to be used as a solar park at the end of the 28-year lease period, and, again, equally revealing, the response was that no one knows what renewable energy technology would be in use a quarter of a century into the future. The solar park model for electricity generation may or may not be superseded. The rapid process of technological and institutional change that led to the Pavagada solar park – fully operational four years after it was announced – points to the rapid and dynamic process that is already in play.

These dynamics were clear to us in conversation with KREDL and KSPDCL in 2022, just four years after our initial visit to Karnataka in 2018. In 2022, we interviewed several of the senior executives in KREDL and attended the KREDL stand at the 'Invest in Karnataka' state trade fair that was held during our time there. The exhibition sought to announce Karnataka as 'open for business', with several hundred local businesses and state employers vying for the attention of possible investors. At the exhibition, we met with KREDL and KSPDCL officials who were announcing plans for a new 5 GW renewable energy project (coincidentally) to be located in the Shimoga district to the North. The new project would be a hybrid solar and wind project with storage, organised on a variant of the Pavagada 'plug and play' model to attract international developers. There was discussion about the lessons learnt from Pavagada, and the need, for instance, to consider agro-solar to enable coexistence with agriculture and also for self-cleaning panels to reduce water use. There was discussion of how to adapt the Pavagada leasing model as well as the question of impacts on livelihood for landless workers and for smallholders. The proposed up-scaled project would, as noted, be located in a more productive agricultural region with less stratified land holdings. These factors may, over time, have an important impact on how the project is conceived and developed.

Since that time KREDL has made a further major announcement that the Pavagada solar park itself will be more than doubled in size. The ambition is astounding: the KSPDCL is at the time of writing assembling a new 'land bank' of 10,000 acres, extending the reach of the park to 22,000 acres. This will add 3 GW in capacity, bringing the entire park to more than 5 GW, almost double the size of the world's current largest solar park (2.8 GW, Golmud in China). The KSPDCL executive celebrates the benefits for farmers and also for communities with 1,600 now employed at the park and growing spending on a Local Area Development Fund using CSR funds. Press reports describe the transformation of the 'most backward taluk of Karnataka into a picture of progress'; the KSPDCL states landowners were now enthusiastic participants in the project 'motivated by the promise of both financial returns and the positive impact on their communities' (Goudar 2023).

Announcements that the new hybrid solar-wind-storage hybrid renewable energy park in Shimoga, and plans for the extension of the Pavagada solar park, would replicate the successes of the Pavagada model obscure a process of social learning through contestation that is taking place in Karnataka's energy transition. The Pavagada model, with its foregrounding issues of land, labour, and livelihood in its leasing model and associated social protection policies, was a response to debates and struggles over earlier stages of renewable energy development, such as in Shimoga, where these issues were largely overlooked in the planning and development process. The highly unequal distribution of economic, social, and ecological costs and benefits from the Pavagada model, and the failures of its safeguard and compensatory mechanism, are feeding into new rounds of struggle over the legitimacy of these new renewable energy developments.

The Karnataka case shows that the legitimacy of renewable energy is a product of constant contestation over dynamics of accumulation, appropriation, and regulation that constitute different development pathways. Impacts are not intrinsic to any particular technology. If it appears that wind farms create less tension or displacement for the people who live with them compared to solar farms, this is embedded in the socio-ecological relations and institutional structures of the different projects and development models. In our initial scoping research before settling on Karnataka, we encountered very similar concerns about lack of compensation and employment opportunities and impact on wildlife among people in villages near wind farms located among the Western Ghats in Pune District, Maharashtra. Importantly, this is a generative process: in Karnataka, community members – both those directly affected by current installations, and those who will be affected by proposed new and extended installations – share their experiences to formulate strategies for how to extract the greatest benefit from existing models of renewable energy development, or to put forward alternative models. Debates about alternative strategies are also reflected in earlier research, which found a parallel process of dispossession arising from the use of ostensibly vacant government lands for renewable energy and argued for 'resettlement and rehabilitation through land-for-land, and not monetary compensation', to improve 'trust in public institutions and foster social acceptance' (Yenneti et al. 2016: 97).

The trajectory of renewable energy development in Karnataka has unfolded over a series of phases in installation, institutionalisation, and regulation. In the case of wind, there were a large number of small-scale wind farms set up in the 1990s, mainly providing local backup power. This was overtaken by a wave of larger-scale wind farms, on a build, operate, and transfer basis, driven by tax incentives. From the 2010s, there were operations established by single owner-operators, generally large international renewable energy companies. In the case of solar power, there was a more rapid jump from almost no solar energy to large-scale solar with the development of the Pavagada power plant. Finally, there is the anticipated emergence of hybrid installations, with wind power as one aspect with solar power and batteries, as well as the expansion of the Pavagada park itself.

The technological shifts were transformations in the socio-ecological relations of energy generated in Karnataka by decarbonising its electricity supply. However, it has not delivered the wider transformation demanded by local communities that could represent a clear break from fossil fuel modes of development by challenging interlocking colonial, gender, and caste-based inequalities and oppressions. As Boyer (with Howe) (2019: 194) has argued, 'it is all too easy for renewable energy development to occur with little or no social, political, or economic transition attached to it'. Yet, as with fossil fuel-based energy production, these processes are always contested (Ghosh 2016), and activists on the ground are acutely aware of what is at stake. As Seeta, an Indian activist for climate democracy we interviewed, put it:

> These stories are not any different from fossil-fuel stories, when we go to coal-bearing states or energy hubs of India. Land-issues, water-issues, livelihood-issues, job-issues etc are present there too. How are we going to come up with the vision statement, looking it … not as a transition but make it into a transformation? (Interview Seeta November 2019)

References

Annigeri, V. & Hanagodumath, S. (2018). *Regional Imbalances in Karnataka Revisited: Reflections on Financing for a Balanced Regional Development*, Dharwad: Centre for Multi-Disciplinary Development Research.

Baker, L. (2022). Procurement, finance and the energy transition: between global processes and territorial realities. *Environment and Planning E: Nature and Space*, 5(4), 1738–1764.

Boyer, D. (2019). *Energopolitics: Wind and Power in the Anthropocene*, Durham and London: Duke University Press.

Bridge, G., Bulkeley, H., Langley, P., & Van Veelan, B. (2020). Pluralizing and problematizing carbon finance. *Progress in Human Geography*, 44(4), 724–742.

Buckley, T. & Shah, K. (2018). *Karnataka's Renewable Energy Transformation*, Sydney: Institute for Energy Economics and Financial Analysis.

Energy World (2018). 'World's largest solar park launched in Karnataka', Energy World, 2 March.

European Commission (2013). *CERs and ERUs Surrendered under EU ETS*, Brussels: European Commission.

Garg, R. & Sundararagavan, S. (2018). *Karnataka Distribution Sector Landscape*, Bangalore: Karnataka Distribution Sector Landscape/Shakti Sustainable Energy Foundation.

Ghosh, D. (2016). 'We don't want to eat coal': development and its discontents in a Chhattisgarh district in India. *Energy Policy*, 99, 252–260.

Gold Standard (2024). '250 MW solar power plant in Pavagada Solar Park in Karnataka', Gold Standard Impact Registry.

Goudar, M. (2023). 'Pavagada Solar Park in Karnataka looks to reclaim 'largest in the world' tag with expansion', *South First*, 19 October 2023.

Government of India (2011). *2011 Census*, Population Data Tumkur District, Table DDW PCA 2917-2011, Household Data Tumkur District, Table HLPCA-29571-2011, Delhi: GoI.

Gurucharan, G. (2022). *Karnataka's Energy Sector: Advancing Second Generation Distribution Reforms, Report of the Committee on Restructuring the Power Sector*, Bangalore: Government of Karnataka.

Hodgson, C. (2021). 'Rush for carbon credits spurs surge in power company schemes', *Financial Times*, 17 April 2021.

Jain, A. (2015). 'Solar energy to bring jobs and prosperity back to parched villages' *World Bank blog*, 22 December 2015.

Kale, S. (2014). *Electrifying India: Regional Political Economies of Development*, Stanford: Stanford University Press.

KERC (2024). *Regulations*, Bangalore: Karnataka Electricity Regulatory Commission.

KREDL (2010). *Annual Report*, Bangalore: Karnataka Renewable Energy Development Limited.

KREDL (2022). *Annual Report*. Bangalore: Karnataka Renewable Energy Development Limited.

KREDL (2019). *Details of Short-Term Tender Notification for the Selection of EPC Contractor for the Implementation of 50MWAC Capacity Solar Power Project at Pavagada Solar Park*, Bangalore: Karnataka Renewable Energy Development Limited.

KREDL (2023). *Commissioned Wind Power Projects in Karnataka, May 2023*, Bangalore: Karnataka Renewable Energy Development Limited.

KSPDCL (2015). *Proceedings of the Meeting Held on 28-09-2015 with DC Tumkur Regarding Fixation of Land Lease Rent*, Bangalore: Karnataka Solar Power Development Corporation Limited.

KSPDCL (2018). *Implementation and Support Agreement for Grid Connected Solar Photo Voltaic Projects for Block No. – 2000 MW Pavagada Ultra Mega Solar Park in Karnataka*, Bangalore: Karnataka Solar Power Development Corporation Limited.

KSPDCL (2020). *Charges Payable to KSPDCL by Solar Power Developers Selected by KREDL (1200MW) and SECI (200MW)*, Bangalore: Karnataka Solar Power Development Corporation Limited.

KSPDCL (no date). *Allocation to ESCOMs from Pavagada Solar Park*, Bangalore: Karnataka Solar Power Development Corporation Limited.

KSPCB (2016). *Consent Order No: CTE-102527*, Bangalore: Karnataka State Pollution Control Board.

Knight Frank (2016). *Social Management Plan: Solar PV Park Nagalmadike Hobli, Pavagada Taluk, Tumkur*, India: Knight Frank.

Knuth, S. (2023). Rentiers of the low-carbon economy? Renewable energy's extractive fiscal geographies. *Environment and Planning A: Economy and Space*, 55(6), 1548–1564.

Michaelowa, A., Censkowsky, P., Espelage, A., Singh, A., Betz, R., Kotsch, R. & Dzukowski, T. (2021). *Volumes and Types of Unused Certified Emission Reductions (CERs)*. Freiberg: Perspectives Climate Group.

Panse, R., Kathuria, V. (2016). Role of policy in deployment of wind energy: evidence across states of India. *Renewable and Sustainable Energy Reviews*, 53, 422–432.

Prabhu, N. (2017). '1st phase of Pavagada solar park to be ready by Dec'. *The Hindu*, 30 September 2017.

PwC (2015). *Solar Park of 2000 MW Capacity in the State of Karnataka, India: Detailed Project Report: DPR*, India: PricewaterhouseCoopers.

Rajendran, S. (2019). Emergence of the largest solar power plant in Karnataka – a microscopic analysis. *Journal of Environmental Nanotechnology, Monitoring and Management*, 8(1), 47–51.

Ramachandra, T., Chandran, M., Bhat, S., Aithal, B., Rao, G., & Mukri, V. (2013). *'Status of Forest in Shimoga District, Karnataka', ENVIS Technical Report 53*, Bangalore: Centre for Ecological Sciences, Indian Institute of Science.

Saldanha, L. & Rao, B. (2020). *Notes for the India Workshop of Governance of SocioTechnical Transformation Project (GoST) ESG Notes for GoST Workshop – January 2020*, Bangalore: Environmental Support Group.

SECI (2010). *Jawaharlal Nehru National Solar Mission: Towards Building Solar India*, Delhi: Solar Energy Corporation of India.

Stock, R. & Sovacool, B. (2023). Left in the dark: colonial racial capitalism and solar energy transitions in India. *Energy Research & Social Science*, 105, 103285.

Stock, R. & Sovacool, B. (2024). Blinded by sunspots: revealing the multidimensional and intersectional inequities of solar energy in India. *Global Environmental Change*, 84, 102796.

Subramanian, M. (2023). India's Quest to Build the World's Largest Solar Farms, The New Yorker, 2 April 2023.

Suzlon (2023). *Group Profile*, Delhi: Suzlon.

UNFCCC (2009). *Project Design Document, Project 3700: 13.75MW MW Wind Power Project in Davangere*, Karnataka, India, Geneva: UNFCCC.

UNFCCC (2017a). *Project Design Document, Project: 3700 13.75 MW Wind Power Project in Davangere, Karnataka, India: Crediting Period Renewal Request*, Geneva: UNFCCC.

UNFCCC (2017b). *Project 3700: 13.75 MW Wind Power Project in Davangere, Karnataka, India, Voluntary Cancellation Certificate VC/095/2017*, Geneva: UNFCCC.

UNFCCC (2020). *Project Design Document: 250 MW Solar Power Plant in Pavagada Solar Park in Karnataka*, Geneva: UNFCCC.

Yenneti, K., Day, R. & Golubchikov, O. (2016). Spatial justice and the land politics of renewables: dispossessing vulnerable communities through solar energy mega-projects. *Geoforum*, 76, 90–99.

Yunker, Z., Rowe, J. & Dempsey, J. (2023). 'Pension Investments in Renewables May Be Undermining a Just Transition', *Jacobin*, 30 October 2023.

4

Brandenburg, Germany

Frank Deutschmann is the mayor of the small community of Heideblick, which lies between Berlin and Dresden on the edge of the Lower Lausitz Heathland Nature Park (*Naturpark Niederlausitzer Heidelandschaft*) and comprises 14 villages with 3,500 inhabitants on about 16,500 hectares. Deutschmann used to work with the railways and then studied to become a treasurer in local government. He has been the mayor of Heideblick since 2012 and experienced the development of wind energy production in his community firsthand. By now, the community has 42 wind turbines with a height of between 100 and 150 metres. The ones closest to the settlements are subject to repowering, which means that they will be replaced with new turbines at least 200 metres high. As Deutschmann says

> We're not going to give wind power a round of applause, just the opposite, because the disadvantages that arise from it are not inconsiderable. It's more force of necessity. A wind suitability area gets declared, where wind turbines can be erected, and our policy is to prevent the worst and put up with what we have to. (Interview Deutschmann June 2020)

Deutschmann has had experience in wind energy production coming to town for years. He has seen various models of ownership and reallocation of taxes, also before becoming mayor during his time as treasurer. He has negotiated with companies and talked to tax collection offices, only to figure out in the end that the taxes that go to his community with the 42 operating wind turbines remain marginal, while large sums flow elsewhere.

> Financially we [the municipality] don't really profit from the licencing fees, they're about 100,000 Euros a year, and at the same time we're saddled with higher electricity prices and all the problems that go with expanding the grid. There are some local residents who become active, others who just resign themselves to what's happening, and a few who are in favour. That's usually those who profit in some way, and I don't hold it against them at all. If they lease their land [for a wind farm] there are very substantial financial rewards, but the community as a whole doesn't get much out of it. (Interview Deutschmann June 2020)

He is, however, not pessimistic or resigned. Instead, he drafted a municipal development contract (*Städtebaulicher Vertrag*) for the upcoming repowering, which includes several aspects of financial benefits for his community, from administration fees to compensation measures (though such contracts have been contested in the past) (Heideblick Municipality 2018). Deutschmann says policymakers have been asleep at the wheel; they have failed to learn any lessons from recent history about what makes people more likely to accept wind farms in their vicinity. He cites the example of Feldheim, a nearby village which has become a model for successful wind power development. The company Energiequelle – the largest developer and owner of wind farms in Brandenburg – negotiated an agreement with the villagers, local government, and the farmers' cooperative which has made Feldheim energy self-sufficient, with its own local grid supplying households directly with electricity and heating (New Energy Forum Feldheim n.d).

If you want people to accept [wind power], then there has to be some advantage in it for them, whether it's cheaper electricity or whatever. You can see examples of where it's worked, like with Feldheim. The level of acceptance is very different there, people look out the window and see the wind turbine, and think, I don't like that, but then they look at their electricity meter and think, hmm, it's not costing me so much, I like that. In that way you really can compensate people, don't you think? (Interview Deutschmann June 2020)

More recently, open-field solar has been gaining traction. Brandenburg offers vast expanses of land with relatively few owners, providing an opportunity for large-scale photovoltaic projects without the complexities of dealing with numerous landowners. Additionally, Brandenburg benefits from an established grid infrastructure, which can accommodate the integration of open-field photovoltaic projects, reducing the need for extensive upgrades or investments in transmission infrastructure. As a consequence, a 'solar rush' (Müller and Pampus 2023) has begun, in which project developers aim to secure land use rights. Deutschmann reports that he receives requests for building 80–100-hectare large solar parks literally every week.

We don't want to prevent them [open-field solar parks]. We only want them to proceed in reasonable fashion, don't we?
INTERVIEWER: So that a wild-west mentality does not develop?
DEUTSCHMANN: Yes, yes.
(Interview Deutschmann June 2020)

Deutschmann calls the current rush for land to install solar parks in Brandenburg 'Wind 2.0', indicating that the modus operandi for the two forms of renewable energy production, especially in relation to planning procedure, is pretty similar.

In this chapter we will examine this assertion, describing and analysing how renewable energy project developers, planning authorities, and local communities argue and operate, to show if they have learned lessons from the expansion of wind

Map 4.1 Brandenburg selected renewable projects (Source: Olivier Rey-Lescure 2024).

energy in Brandenburg (key sites for the research are marked on Map 4.1). Will solar developers simply repeat the mistakes made in wind park installation, leading to an erosion of acceptance and social legitimacy as more and more solar parks appear in the landscape? Or are other forms of development and social engagement with renewables possible?

Brandenburg's Energy Transition

The state government likes to portray Brandenburg in its official publications as an energy state, and pushed ahead with the expansion of renewable energies comparatively early. In 2008, 2010, and 2012, for example, Brandenburg was awarded the guiding star (*Leitstern*) for its pioneering role in the expansion of renewable energies. In 2023, 95% of electricity consumption in Brandenburg was covered by renewable energy, according to the Ministry of Economics, Labour, and Energy (MWAE 2023). Brandenburg has the second highest installed capacity of wind power of any German state, with about 4,000 wind turbines installed generating 8,000 MW, and in 2022 ranked third in installing new net capacity, after Schleswig-Holstein and Niedersachsen (Deutsche WindGuard 2023).

Along with this high concentration of wind, roughly 35,000 photovoltaic systems and over 500 biomass plants, Brandenburg has the highest installed electrical capacity from renewable energies per inhabitant in a nationwide comparison (MWAE 2023), although it needs to be acknowledged that the state is comparatively sparsely populated. But there are also significant conflicts over wind energy in Brandenburg, which have been developing and intensifying over the last decade (Bues 2020). These conflicts are now exerting a growing influence on policymakers and culminated in the declaration of a temporary moratorium on wind farm development in 2019, when the existing planning regime was ruled invalid. The moratorium ended in November 2022 with the passing of a federal law to accelerate the development of wind energy (Windmesse 2022). Understanding these conflicts, and how and why they arise, is all the more important because, as outlined in Chapter 2, the future of Germany's energy transition or *Energiewende*, and its emission reduction goals are at a crucial juncture.

There is extensive research on conflicts over wind energy in Germany, analysing them from a number of disciplinary perspectives. Economic and technical analyses abound, as do policy recommendations for fostering the energy transition. Spatial analyses as well as psychological ones on the impact of wind turbines exist as well, as do a small but growing number of studies based in detailed local ethnographies (Hübner 2012; Leibenath et al. 2016; Gailing et al. 2019; Bues 2020). There is less research on solar energy, and studies which compare and contrast the contestation and social acceptance of wind and solar energy production at a local level are thus far largely missing. This is surprising, given the acknowledged need to stress social acceptance in situ rather than just nationwide (Löschel et al. 2021; see also Wolf 2020) and the comparatively long life of the German energy transition, especially.

To fill this gap, we have been conducting ethnographic field research in the state of Brandenburg – as mentioned, one of the German states which has sought to lead the way in implementing the Energiewende, but as an East German state with a long history in coal mining also one with its own particular features (see Müller and Morton 2021a). This chapter thus rests upon ethnographic field research in Southern Brandenburg, mostly in the region of Teltow-Fläming, since 2018, comprising participant observations and deep hanging out (Geertz 1998) in villages and small towns where solar and wind installations took place, are planned or in progress. It is thus an ethnography of the German renewable energy installation, which aims at a holistic view, that is, understanding and analysing the emic perspectives of most if not all actors involved in the processes of developing particular wind and solar parks. We participated in and observed council meetings, public hearings, formal and informal gatherings, village fetes, and simply everyday life. We also conducted and recorded more than two dozen formal interviews, albeit acknowledging the difference between carefully worded statements in recorded interviews and informal conversations and people's actions (see also Müller and Morton 2018).

In our analysis, akin to the approaches taken in the Indian and Australian chapters, we adopt Becker and Naumann's (2018) typology of energy conflicts to discern similarities and distinctions between solar and wind energy production within the German context. Their typology, grounded in empirical studies of local energy conflicts across Germany, offers a fitting framework for our investigation. As evidenced in the ensuing case study material, we encounter the same categories of conflict and configurations of stakeholders outlined by Becker and Naumann. In elucidating the backdrop to these conflicts, in this chapter we additionally draw upon the concept of assemblages – reverberating the acknowledgement of non-human actors in prior studies of wind energy politics in Germany (Gailing and Moss 2016; Chezel and Labussière 2018) and relating to assemblage theory as used more broadly in anthropology (Marcus and Saka 2006; Helfer 2022). Objects and artefacts in socio-technical systems can actively shape how an energy transition takes place, also beyond human agency (Gailing and Moss 2016: 111–112). An example of this might be the increasing height of wind turbines, which gradually over time may shift the attitudes of people who live near them from acceptance to resistance; in other words, the actual configuration of the objects (turbines, solar arrays) and their arrangement in the landscape can contribute to legitimising or delegitimising the local development of wind and solar energy.

Viewing energy transitions through the conceptual lens of assemblage theory stresses the interrelatedness of different human and non-human actors and allows us to navigate the notion of an 'anti-structural' process, characterised by emergence, heterogeneity, and decentralisation within a structured social context (Marcus and Saka 2006: 101). Energy transitions, we argue, are intertwined within complex social constellations that extend beyond electricity generation and predate the emergence of modern renewable energy technologies. Viewing energy regions and their inhabitants as integral components of a long-evolving assemblage involving technology, human and non-human actors, landscapes, and environments enables us to integrate the insight that landscapes are not 'static "scenery" for displaying pre-disposed attitudes'; they are cultural objects which can be mobilised discursively in the social making of arguments (Reusswig et al. 2016: 226).

Our approach, however, remains firmly grounded in the framework outlined in Chapter 1: thus, while technological and natural 'objects' (wind turbines, forests) may exercise a certain kind of agency, we emphasise the forms of human agency which have placed the turbines in the landscape in the first place; the competition (rivalry) between turbine manufacturers to produce larger and more efficient turbines; the needs of investors and project managers to maximise profit; in other words, the specific configuration of neoliberal energy capital itself and the state policies (regulation) which enable it. Through our case study material, we uncover the nuanced perspectives of various stakeholders involved in the localised energy transition, highlighting both the advantages and drawbacks perceived over time. We track the evolution of wind energy planning and production as precursors to the current proliferation of land acquisition for solar parks. Additionally, we illustrate how shifts in energy governance and community engagement significantly influence attitudes towards wind and solar parks.

Wind Park Planning I – 'Gradually, You Start to Feel Surrounded'

On a summer evening in June 2019, we sit on the garden terrace of a house in the village of Niebendorf-Heinsdorf. The house belongs to Stephanie, one of the founding members of the citizens' initiative 'Country Life' (*Landleben*). Stephanie describes herself and her husband as 'dyed-in-the-wool' Berliners who moved to the region after reunification. They bought an old manor house in 1993, renovated it themselves, and moved in in 1995, later bringing Stephanie's parents to live with them in the village. Both Stephanie and her husband still travel 70 kilometres every day to and from Berlin, where they work for a federal government authority. They originally moved to the village because some friends from Berlin had done so too, but chose to stay after their friends moved away because they wanted to bring up their children in the country, and because their house looks out on forest, meadows, and fields.

Stephanie says she first started to become aware of wind turbines in the landscape around the year 2000:

> I think it was around the beginning of 2000 when the first wind turbines began to appear in the area around Hohenseefeld, although we only really realized it was happening when the towers were there all of a sudden. No one had spoken to us about it beforehand, at least not here in our municipality, or in the local consultative assembly, nothing was said, and so it was more or less something which just crept up on us. (Interview Stephanie 2019)

More and more turbines began to appear closer and closer to her village, totalling 33 in 2019, with more in the planning. Stephanie and a few other engaged citizens learned pretty quickly that large-scale public protest against the increasing number of turbines, and the feeling of being surrounded, would hardly be possible in a rural area like theirs. They rather took to informing themselves about the administrative processes surrounding wind development, means of intervening, and how to access information. These administrative processes involve different levels of government, and understanding how they function and interact is crucial to understanding the overall governance of renewable energy in Brandenburg.

The Governance of Wind Energy in Brandenburg

The early development of wind energy in Brandenburg in the 1990s was relatively unregulated. Permits were granted to individuals or companies planning to install wind turbines by local authorities, which generally preferred them to be located in outlying areas. Regional planning was introduced in 1993 to regulate land use. In 1997, wind energy was privileged over other forms of land use in Brandenburg, and in 2003, a federal law was passed which designates wind suitability areas (*Windeignungsgebiete*; Federal Parliament 2022). Despite their name, these wind suitability areas are not necessarily in locations with the strongest or most reliable wind. They are determined by geographical planning, which takes into account social, economic, and ecological considerations as well as the technical needs of wind energy production. Regional planning bodies govern on-the-ground development of wind energy in Brandenburg and are required to balance all of these factors when making decisions.

The state of Brandenburg is divided into five regions for this purpose. Each region has a regional planning body (*Regionalplanungsstelle*). Geographers working in these bodies map and plan the region, designating areas for wind energy production and also for farming or flood control. The maps they create and publish are the legal basis for all construction activities and approvals in Brandenburg. Wind planning within these maps takes into account so-called hard and 'soft' criteria, for example, environmental protection or distance to buildings. Once these criteria have been applied, the remaining areas of land available for wind farms are dispersed and fragmented and amount to just over 2% of the total land surface area of Brandenburg.

As Müller and Morton (2021b) have argued, these regional planning bodies are now the principal vehicle through which the energy transition is governed and regulated. They are a way of taking the social dimension and 'human drivers' of the energy transition into account, allowing for 'conscious governance' and the involvement of a wide variety of actors (Renn 2011; Kern & Rogge 2016). Regional planning bodies also try to tackle the question of the social legitimacy of renewable energy production. This can involve political delegates of larger communities as well as (on application) representatives from unions, environmental

NGOs, and other civic stakeholders in efforts at raising 'social acceptance', so that those involved in and affected by the process of energy transition feel they have a stake in the process, or the outcome of installing wind plants.

This, at least, is how the process is supposed to work. What local residents affected by wind development experience on the ground, however, may be very different. When Stephanie, the resident of Niebendorf-Heinsdorf mentioned earlier, started to become concerned about the wind turbines appearing in the landscape around her village, she contacted the local authority in Dahme (*Amt Dahme*), the local government area within which her village falls. She was told that the wind farms were in an adjoining local government area, where the Dahme local authority had no jurisdiction.

The local advisory committee for her village (*Ortsbeirat*) had not been informed about the wind farms, neither by the authorities in Dahme nor by the adjoining authorities. Even within the formally affected area, only local landowners were approached directly by project developer(s) wanting to lease their land for wind installations, but according to Stephanie, residents of the villages who do not own land were left in the dark. In order to be closer to information about what was planned, Stephanie got herself elected to the village advisory committee in 2009. From then on, she began to inform herself about the planning processes at the regional level, the options a community has when wind suitability areas are declared, and the individual options to intervene at hearings and during the public review processes. She also reported back to the committee about proposed wind developments that she now came to know about at various planning stages.

Together with a resident of a neighbouring village, Holger, Stephanie founded the citizens' initiative Country Life (*Landleben*). They ran candidates for the local council in the municipality of Dahme (*Stadtverordnetenversammlung*) and both Holger and Stephanie were elected (2 out of 18 seats on council), further fostering their access to information and participation in official procedures. Stephanie and Holger continue to engage with the energy planning process from a critical stance, making numerous formal requests and contributions to the debate about wind energy development in their area. They achieved not only a deep knowledge about how and when to question and intervene in wind turbine construction plans but were also successful in influencing local benefits and locational planning, including putting a halt to half a dozen turbines in the area.

How Energy Conflicts Undermine Legitimacy: Wind Park Planning I

Neither Stephanie nor Holger, or indeed the citizens' initiative they founded, express any fundamental opposition to wind energy as a form of energy generation. This reflects a broader tendency identified in previous research: peoples' attitudes towards renewable energy, and wind in particular, are not predetermined, but are formed over time and shaped by the dynamics of local conflict. They are 'emergent, negotiated and shifting' and landscapes such as the countryside of Teltow-Fläming become arenas for the social making of arguments, instead of simply the static 'scenery' against which pre-disposed attitudes are played out (Futrell 2003: 360; Reusswig et al. 2016: 226). While the bureaucratic processes of energy governance described in the previous section are intended to enable social co-existence between wind energy and other forms of land use – an integration of wind energy into the fabric of everyday life – their effect in practice has been to undermine that coexistence (Müller and Morton 2021b).

In this particular instance, several of the types of energy conflicts and legitimacy problems we identify in Chapter 1 are visible. The underlying difficulties of legitimising the appropriation of wind power in terms of the types of technologies that are deployed and the locational issues they create are clearly in evidence. It is equally difficult to establish public support for the process of accumulation, especially as it centres on private corporate players, and only some existing landholders. With problems of technological, locational, and distributional legitimation, regional authorities can struggle to be seen to be acting in the public interest. Regulation itself becomes a site of contestation, in terms of the degree to which state authority privileges some and marginalises others. Procedural legitimacy is thereby put into question, along with any effort by the regional state at charting a consensual renewable identity for the region to guide future planning.

We can see these aspects in play in opposition to the local development of wind energy as expressed by the citizens' initiative and its members. There is a clear perception of a distributional conflict, revolving around the ownership of land, and who benefits from the financial rewards of wind installations. Stephanie points towards the logics of financial accumulation at play, where landowners are able to capture new forms of rent from renewable energy production. She says she can understand why some local landowners chose to lease their land to win project developers: 'if someone says, well, I can get 30,000 Euros a year for a hectare of land if I lease it, for some people that means they wouldn't have to work anymore'. But she says undue pressure was put on some local residents by project developers to lease their land, and some of the land was leased to developers by absentee landholders, who live far away and are not affected by the developments. Some locals have leased their land to large agricultural companies who in turn may lease it on to wind developers. According to Stephanie, there are few residents who have refused outright to allow wind turbines on their land.

Both Stephanie and Holger stress the lack of access to information and opportunities for participation and transparency in the regulatory processes instituted by local authorities, which are common features of procedural energy conflicts. Both experienced the difficulty of finding out what was proposed and how far projects had progressed through the approvals process. It was only when they became involved in local government, through the advisory committees of their respective villages and as members of the municipality's council, that they started to become aware of issues and means to steer and influence land use and regional plans, development processes, and approval procedures. At the same time, they grew increasingly aware of locational conflicts: the impact of wind turbines on land use and the appearance of the landscape. At the root of this conflict is a contradiction between, on the one hand, corporate imperatives to maximise the appropriation of land for wind energy and, on the other, the use values of the existing landscape for local residents. This was a gradual process, something which, as Stephanie put it, 'more or less crept up on us' (Stephanie, Interview 2019). She could have lived with the 33 turbines which began appearing south of her own village around 2000, she says, but when a further 20 wind turbines were installed north of her village, and then further ones were proposed for to the south-east and south, only 1 kilometre away, 'gradually, you start to feel surrounded' (Stephanie, Interview 2019). For Holger, it was seeing the larger picture in the regional plan that really brought home what was happening:

> you could see from the plan that Dahme was going to be surrounded, but the straw that broke the camel's back was when I saw a land use plan for the village of Rietdorf, a kilometer away from Gebersdorf, proposing 56 wind turbines with a height of 250 metres. (Interview Holger 2019).

It was this perception of a gradual encroachment on the landscape which led them to found the citizens' initiative Country Life and to run as candidates in the local council elections. 'We aren't militant opponents of wind energy', says Holger, 'and that's not the way we wanted to present ourselves. We didn't want to say, as an election slogan, the only solution is no windmills, but rather that there has to be harmony' (Interview Holger 2019). Holger and Stephanie support options such as the so-called 10 H rule, which governs wind development in Bavaria and stipulates that there has to be a distance of 10 times the height of a turbine between it and the nearest settlement. Here an important feature of recent wind development comes into play: technical innovation has enabled taller and taller turbines to be built and installed, often (but not always) as part of the process of repowering, which is the replacement of an old wind turbine by a new one on an existing site. The regulatory regime governing wind energy planning has failed to keep pace with this development; decision-making bodies plan and map a wind suitability area onto the landscape without taking the third dimension – height – into account (Müller and Morton 2021b).

This third regulatory dimension can influence the attitudes of local residents who would otherwise welcome wind projects. Christoph, a farmer whose farm lies about 10 kilometres from the villages where Stephanie and Holger live, says that in principle, he'd be happy to say yes to wind turbines on his land: 'it would mean less work on the fields, and I'd be less dependent on the crop and its market value' (Christoph interview, 2019). Christoph's farm covers 180 hectares, and when we spoke to him in 2019, there were no wind turbines on his land, as it was not designated as part of a wind suitability area. A few months later, in early 2020, the situation had changed slightly. A wind project developer had approached him, offering to lease the small patch of forest he owns. The project developer was rightly speculating that a new wind suitability area would be designated nearby, including the forest that Christoph owns. The developer wanted to construct a wind farm of up to 21 wind turbines; a cable or substation would be likely to stand on or pass through his forest. But the catch was that the wind turbines would be up to 240 metres in height, and only 1 kilometre from the nearest village, where his parents and some of his other relatives live. Christoph was not sure if that wasn't too close for them, too much of a change for the location, and was consequently not so sure about signing the lease.

These locational conflicts, increasingly expressed in three dimensions, overlap with and shade into what Becker and Naumann categorise as identity conflicts, which affect the perception of a region and its distinctive identity by locals and outsiders. These identity conflicts are often bound up with conceptions of nature and rurality (Becker and Naumann 2018: 515). For Stephanie and other members of the citizens' initiative, the forest is central to what they perceive as the rural identity of Teltow-Fläming and its landscape: As Stephanie argues:

> There shouldn't be wind turbines in the forest, because climate protection shouldn't come at the cost of nature conservation. Trees are being cut down to make way for wind turbines. Some people say, well, it's just pine forest, it's there for forestry, we don't need every tree. But I say, of course we need every tree, because the forest is a habitat for all kinds of animals, birds in particular. The most important thing is that we don't destroy nature to protect the climate.
>
> INTERVIEWER: But isn't protecting the climate part of protecting nature?
>
> STEPHANIE: It should be, but surely that shouldn't mean destroying what preserves my climate right here, and that's the forest.
>
> (Interview Stephanie 2019)

Holger supports Stephanie, arguing that on the one hand Germany is trying to extend forest coverage, and on the other promoting wind development, which makes significant inroads into the forest. Their views are echoed by Cindy, who is in her early thirties, and together with her husband runs a livestock and crop farm in the nearby village of Wahlsdorf. They are currently in the process of shifting from conventional to organic farming, and thus to something they think is more sustainable. Nonetheless, Cindy also opposes any further expansion of wind energy.

> Well, it does disturb me, that the way the landscape looks is being destroyed [by wind turbines]. I thought I'd get used to them with time, and I try not to see them or take notice of them at all, but of course, that's difficult, just because they're relatively large, and not so easy to overlook (Cindy, Interview 2019)

And Cindy, who is a keen horse rider, adds

> We often ride through the forest, and it's not the same now as it used to be. Five years ago, when we went riding there, it was all still green and natural, and today it feels like every hundred and fifty metres you find a wind turbine in between the trees. (ibid).

Landscapes, and 'nature' itself, are socially constructed (Cronon 1995: 25); what Cindy and her neighbours perceive as 'natural' in the Brandenburg landscape has, in fact, been altered and shaped by human action over thousands of years (Küster 2010). But while the landscape that Cindy and other members of the citizens' initiative believe is being 'destroyed' may be a human construct, it is central to what they see as the identity of the region. As Reusswig et al. argue, the landscape is not 'static "scenery" for displaying pre-disposed attitudes', but is in a process of dynamic transformation, as the increasing density and height of wind turbines make them impossible to overlook (Reusswig et al. 2016: 226). As the drive to make wind energy profitable and investable leads to taller and more socially and spatially proximate turbines, contestation is largely not focused on support for or opposition to renewable energy or action on climate change. Rather, contestation over wind power (and indeed solar energy) coalesces around the patterns of appropriation, in particular land and surrounding ecosystems, that make renewable electricity possible.

It is, eventually, often nature and rural identity that are mobilised strategically to contest the construction of wind farms, with the intention of slowing down development or stopping it altogether. These contests are primarily mediated by state planning authorities and processes. In 2015, two different project developers applied for approval to install 10 turbines each in the Rietdorf/Illmersdorf forest, at a location 1 kilometre away from Stephanie's village. The citizens' initiative conducted observations of birds in the forest and was able to show that a pair of owls were nesting there. As a result, the state environment authority withdrew approval for half the area the project developers had applied for and subsequently withdrew approval for the remaining half when the citizens' initiative provided evidence that red kites were also nesting in the forest. The citizens' initiative had prevented the installation of the 20 turbines. Their attentiveness and intervention forced the companies to apply for only five turbines, which were eventually approved in 2020 and built in 2023. In 2019, Stephanie described it like this:

> The approval hasn't been given yet, and we've been waiting for more than a year for a construction plan (*Bebauungsplan*). We asked to see the construction plan at the local authority in Dahme, but we get the impression no-one's taking any trouble to respond to our request. If the main application is approved and there's still no construction plan then the wind park can be extended as much as the project developer likes. But the fact that the main application hasn't been approved is thanks to the people from our citizens' initiative who've been observing the populations of owls, red kites and black kites in the forest and passing that information on to the State Environment Office. (Interview Stephanie 2019)

As we can see, a number of different types of energy conflict and constellations of actors overlap in this particular location. There is fundamental distributional conflict, revolving around the ownership of land and who receives economic benefits from wind development. Procedural conflict reflects the perception of local actors such as Stephanie and Holger that local residents aren't informed about wind development unless they happen to be landowners; in other words, there is a lack of procedural mechanisms which automatically involve residents in the local governance of wind energy.

As Becker and Naumann argue, we can see these procedural conflicts as also having a productive dimension. The distributed nature of wind energy generation has brought with it the evolution of a distributed system of energy governance. The various opportunities for public participation opened up by regional planning of wind energy allow for greater citizen participation in the governance of the energy system than was the case with coal-fired power – especially in Eastern Germany. Dissatisfied with not being heard and with the limited opportunities to influence wind development through public hearings, local residents like Stephanie and Holger have acquired expertise in the planning, financial, environmental, and technical aspects of wind energy. They have learned how to intervene in planning processes, for example, through monitoring wildlife and reporting these findings to environmental authorities, and consequently influencing how wind energy development is governed. The citizen's initiative expanded its tactics from attending public consultations and petitioning the state parliament, to influencing wind development through being informed, informing others, and actively participating in local development and governance. These strategies, in turn, create political pressures for both authorities and developers to incorporate new actors and concerns within planning processes and investment plans.

Locational and identity conflicts overlap and have been intensified by the allocation of wind suitability areas, which was intended to confine wind development and has helped to establish and enforce environmental protection, but has actually had the effect of concentrating the burden and potential negative impacts of wind energy in particular locations. It has resulted in increasing spatial concentration of wind energy, higher density in those areas designated for wind, and an expansion upwards, into increased height (Müller and Morton 2021b) – examples of what Gailing et al. (2019) summarised as the *Energiewende*'s wide-ranging impacts on the spatial structures of the energy sector as it plays out on the ground. While some regions in Brandenburg have close to no wind turbines at all, the community of Dahmetal (comprising of three villages on 45 hectares), for example, has 45 of them installed, taking up 21% of the community's land. Dynamics of accumulation and appropriation in Brandenburg's wind energy industry have thus produced a landscape of uneven renewable energy development.

As we have argued elsewhere, the increasing density of wind farms and the increased height of the turbines themselves undermine wind energy's advantages over other forms of energy generation (Müller and Morton 2021b). Wind energy can co-exist with other forms of land use, such as farming; crops can be grown underneath the turbines and stock can graze around them. But the increasing concentration and height of turbines leads people to perceive them as a threat and encroachment, rather than something with which they can co-exist. The height of turbines plays a central role in how residents perceive their surroundings in terms of views from windows, when driving by wind parks on their way home, or walking or riding in the forest.

As Schöbel and colleagues (2022) have shown, people living in a region would equally distribute wind turbines across the whole area, if given the chance to decide. They would intentionally factor in the potential disturbing effects of wind turbines and try to distribute them more evenly. What Schöbel and colleagues argue against on psychological grounds, Boyer and Howe (2019) do on grounds of politics and justice: the concentration of wind turbines in ever larger parks and its international or non-local financing produce unequal and unjust constellations. The majority of revenues flow to investors outside a region, while the impacts (on fishing, food production, common law, customary access to land) happen within the region or locality. Once again, the dynamics of appropriation and accumulation tend to undermine the social legitimacy of renewable energy at a local level.

Stephanie also puts the concentration in her region in words, saying she'd be prepared to accept more wind development in her immediate surroundings if the burdens were more equally distributed between country and city:

> I wouldn't like to see it all happening in one place [the expansion of renewable energy], but rather everyone doing their part. Then the wind farms would have to be closer to the cities. When you look around the outskirts of Berlin you don't see any now. They're all a considerable distance from the city. Probably they [the people in power] know very well that if they started building them closer to Berlin or Potsdam there'd be protests, the people there would start making a noise. So I'd be prepared to put up with two hundred wind turbines in front of my door here, but only if they're also in the other Federal states as well, thank you very much, and spread out evenly as far as that's possible. (Interview Stephanie 2019)

The interaction between these dimensions of legitimacy, in terms of the extent to which they produce a reinforcing or cumulative dynamic, can be vitally important in shaping (or indeed pre-shaping) the agency of the various actors. The field of legitimation and de-legitimation can, as discussed in Chapter 1, be analysed as an 'assemblage of agency' put into play by the transition process. This has contingent and autonomous 'moving parts' that nonetheless are correlated in reflecting the underlying dynamics of appropriation, accumulation, and regulation, albeit in a more synthesised way offering resonance across multiple fields at once. Regional planning is one instance of this where very often there are deliberate efforts at achieving regional agglomerations for the emerging industries.

There are multiple dimensions of energy conflict. These conflicts extend beyond the mere financial benefits allocated to stakeholders in energy production, encompassing the unequal spatial placement of wind turbines within regions. This adds layers to the complex assemblage involving wind turbines, landscapes, human communities, and biodiversity. Particularly significant is the transformation of rural areas, previously characterised by agriculture and recreation, into industrialised landscapes, leading to tensions and resentment among the local population. This unequal allocation of wind turbines, notably in rural Brandenburg, compared to electricity consumption centres like Berlin or industrial hubs in Southern Germany, exacerbates the legitimacy challenges facing the energy transition. Distributional conflicts do not only occur over the surplus generated by renewable energy production but also where renewable energy accumulation takes place and who gets to decide. The resulting discord potentially undermines the social acceptance crucial for its continued advancement.

Recognising this decline in legitimacy, both the national and Brandenburg state governments, along with project developers, are actively seeking ways to address these concerns, underscoring the evolving and dynamic nature of the energy transition process. The following example will illustrate how they are approaching the growing apprehension about renewable energy development.

Wind Park Planning II – 'It's Strategic'

In summer 2019, the wind energy company Prokon invited all the owners of forested land in the vicinity of Welsickendorf, a small village in southern Brandenburg, to a meeting in the local community house, a low-rise building spreading the charm of a functional flat roof building erected in the 1970s. It was sticky and gloomy inside on this hot summer day, and Prokon's regional office manager and the field manager had set up the laptop and beamer and brought brochures, pretzels, and beer. They welcomed about 30 participants and after briefly introducing the company, a co-operative with 3,000 members and 200 employees based in Germany's north-west, introduced the precise planning for a wind park of 21 wind turbines planned to be erected about 1 kilometre south of the village. The turbines, explained the Prokon representatives, would be about 240 metres in height, each requiring a concrete foundation of 20 metres in diameter and about 4 metres deep. They would have a capacity of 4.2 MW, and with a price of 5.25 cents per kWh generated, the company would pay a total of about 1 million Euros in rent for all 21 proposed locations where the turbines would be sited. The unusual aspect was that Prokon offered a 20:80 model of how the rent would be paid: 20% of the rent would be paid directly to those owners who actually own the plots of land where the wind turbines were to be erected, and 80% of the rent would be shared among everyone who owns land in the whole designated area, according to what proportion of that area they own.

At the meeting the field manager also showed the timeline for the wind park: while Prokon were currently in the process of acquiring lease contracts with owners, he explained, they would do first environmental assessments and prepare the required documents for permission in 2021/2022, undertake final planning in 2022, and ideally hand in the documents and receive the necessary permissions in 2023. Reverse auctioning would take place in 2024 and construction could start, making 2025 the year for bringing the wind turbines into service.

Prokon is a cooperatively owned renewable energy developer and operator, focused on wind energy. It claims to be the largest energy cooperative in Germany, with around 40,000 members. As of 2023, Prokon had developed and has ownership stakes in 57 wind farms with an installed capacity of 660 MW in Germany, including 10 wind farms in Brandenburg, as well as others in Finland and Poland. Prokon itself has a tumultuous history that led to its re-launch as a cooperative. The company was originally established in 1995 as a for-profit business. It utilised an unconventional financing strategy in which small investors would purchase rights to dividends without ownership. Concern about the financial viability of the model led to investors withdrawing their capital and Prokon was forced to file for insolvency in 2014 in a high-profile case that brought public attention to both renewable energy investment models and gaps in financial regulation. Declining a below-par offer by big German energy utility EnBW, the remaining investors instead voted, by a margin of 90–10, to transform Prokon into a cooperatively owned company in 2015. It was reported that Prokon's investors-turned cooperative owners were motivated by 'being a living part of the citizen-driven *Energiewende*' (DW 2023: 1).

The local government area where Prokon wanted to build near Welsickendorf already had almost 5% of its land allocated to wind energy production. Questions arose from among the audience, because the proposed site was not a designated wind suitability area. By now, in 2019, the moratorium was in place, effectively forbidding the installation of new wind turbines in Brandenburg – hence there was currently no way for the development to proceed. This was true, answered the field and office manager. He acknowledged that there was a speculative aspect to the proposal Prokon was putting before the meeting; however, it was based on a strong expectation that the site would be designated as a wind suitability area in the future (this approach is known as white planning; see Müller and Morton 2021b).

It is unusual for a wind park developer to inform (parts of) the public at this comparatively early stage. German law requires public hearings only at a very late planning stage – leaving little to no room for alternations or amendments. Early meetings, by contrast, carry the risk of, and potential for, stakeholders demanding changes. This is precisely what the landowners invited to the meeting in Welsickendorf did. They demanded alternations to the contracts and more protection for the trees they are caring for and harvesting. But project developers like Prokon had learned from experience that it was much more advantageous to involve the public early on in the development process and make alterations and concessions, rather than leave the publicising of proposals to the final stages. At the meeting in the village's community house in 2019, people voiced little criticism.

Two years later, in March 2021, we conducted a poll, showing that more than half of the inhabitants (not all of them forest owners) supported the construction of the wind park, or could accept it with conditions. The poll was conducted in the village of Welsickendorf in March and April 2021 in the form of a written questionnaire comprising boxes to tick, scales, and open remark fields. We requested the 200 inhabitants in 86 households to participate. Twenty-six households responded through filling out and sending back the questionnaire, making the poll a representative study. We asked if people see advantages and/or disadvantages in wind energy in general, what those would be, in what fields and sectors a negative/positive impact is anticipated, and ended the questionnaire with asking about the attitude to the precise planned wind park near Welsickendorf. Given this, it is unlikely that protests will emerge in the future concerning this project.

During the same time (between 2019 and 2021), a number of political measures were taken to increase the legitimacy of wind installation in rural Brandenburg. One example at the regional level is the lowering of the threshold for participating in regional planning: while before only representatives of communities or local government areas with more than 10,000 inhabitants were allowed a seat at the regional planning body, those with 5,000 inhabitants could now participate (including now, for example, also a representative from the *Amt Dahme*, which is responsible for both Stephanie's village and Welsickendorf, where Prokon wants to operate). The two-year moratorium on the construction of new wind turbines, enacted at the state level by the Brandenburg government in 2019, was implemented to ensure the redrafting of regional plans after courts declared them void (thereby effectively ruling in favour of wind park developers), but arguably also as a reaction to increasing conflicts over wind development. In 2019, the Brandenburg government passed the Wind Energy Levy Act requiring the companies to pay 10,000 Euro per year for every newly constructed wind turbine to the communities within a radius of 3 kilometres from the turbine (Brandenburg State Parliament 2019). In 2021, the federal government followed suit with an amendment to the renewable energy law, requiring companies to pay 0.2 cent per kWh generated wind energy to the affected communities – with the government reimbursing the costs to the companies. The companies and communities are hence very likely to make use of this option.

The Case's Energy Conflicts

This is not to say that these financial benefits for the communities and the increased opportunities for the involvement of the public and local government, are pacifying all energy conflicts, or that they offer an ultimate solution for generating social legitimacy. Prokon's efforts to address distributional and procedural conflicts are having the effect of channelling and managing concerns, rather than resolving them. Prokon tries to mitigate procedural conflicts through informing and involving local residents before the formal process of seeking approval for the project has begun. The company also informs the local administration at a quite early stage, even though their approvals are necessary only after several surveys and expert reports (requiring several years). Prokon consciously and carefully sets out to see first if a substantial number of owners would be willing to participate, choosing the right time to inform landowners, the whole community, administrative bodies, planning bodies, and other political decision-makers. These strategies, which to an extent socialise Prokon's investment plans within local communities and authorities, attempt to build legitimacy with citizens who have previously demonstrated their ability to influence outcomes for developers. Their approach is to inform and include, positively influencing stakeholders, and in the process keeping competitors and protesters at bay.

This course of action can be seen as a form of learning from previous conflict and experimentation with new forms of energy 'governance' which are not imposed from above by federal, state, or local government but which grow out of the processes of contestation and the participation of citizens (as Becker and Naumann (2018) also argue). This mitigation of procedural conflicts, however, is also seen critically by local villagers. Ulrich, a retired teacher and school director and active in local parliament since the 1980s, draws on experiences with earlier wind developers when judging the current project. He recalls how 'men with suitcases walked around the premises with recording devices when people gathered for local council meetings, in order to find out who else they could influence'.

Once, he tells, he was called by someone from a wind development company in Bremen, which invited him and his wife to dinner in Berlin.

> There was not much about dining, but he told us that he has six lawyers sitting there and they will build the wind park with the community, which then will profit from it, too, or without it, and then it will gain nothing. And I told myself: 'Well, now definitely not. You don't do things like this.' And he was right after all, it all happened as he said: They built the wind park and we got nothing out of it. (Interview Ulrich August 2019)

Against this backdrop, Ulrich has more recently been dealing with Prokon, including face-to-face discussions with the field manager, and assesses their methods as more fair. But he still emphasises that the company's approach is primarily a strategy geared to securing consent:

> Strategically, they are doing it exactly right: they convince the owners, then take them along to council meetings and the like, where they also tell their local government representatives that they want this [wind park], and eventually the representatives of the local council are in the position to alter the local land use plan [in favour of the wind park, if no regional plan is in place]. … And [the field manager] is also clever. It is like strategic warfare. … And once the wind turbines are standing here, they stand here for twenty years. And then we don't know how planning will continue, if, one day, we will be surrounded (Interview Ulrich August 2019)

Hence, the reduction of procedural conflicts is not the result of an altruistic approach or intrinsic energy justice. The changes in information flow and the more consensus-building approach rather stem from wind park developers' experience in how to build wind turbines in a way that minimises conflicts and resistance, which for the company would always require more resources.

Prokon's 20:80 benefit-sharing model has also emerged from their experience, learning from previous energy conflicts, and as with the abovementioned legislative changes, aims at lowering distributional conflicts and addressing spatial divides. Principally, Prokon seeks to share income more equitably among those who own land on which turbines will be sited and those who do not. It is an example of *re*distributing profits from renewable energy accumulation but is not a *pre*-distributional approach that would fundamentally alter social relations of energy production. Despite being cooperatively owned at the enterprise level, Prokon does not offer profit-sharing or co-ownership for local communities. Benefit-sharing instead effectively insulates profit flow for the company, protecting it from further local claims. It signally fails to address the central distributional injustice relating to the mainly corporate ownership structure for renewable energy and the extent of the profit taken from renewable energy in Brandenburg, which will roughly amount to 50 to 70 million Euros in 20 years for a 20-turbine wind park like the one planned in Welsickendorf (see details later). Arguably, to maximise public benefit from the energy transition, renewable energy needs to be recognised as a public good and its ownership structure needs to reflect this.

When talking about the German energy transition as a form of democratisation (Jungjohann and Morris 2016), contemporary Brandenburg wind energy systems are not a good example. Imperatives of large-scale investment dominate, leading to little-to-no profit for those living in close proximity to the power plants. The reasons for this are to be found in the development of the Renewable Energy Law and the concentration of wind parks (see Müller and Morton 2021b), as well as in the distribution of financial capital and expertise. A wind park of the size Prokon plans requires several surveys and reports, plans, modellings, permits, and permissions, which will cost about 5–6 million Euro. Actual construction costs are calculated at about 5 million Euro per turbine. The whole planning procedure will take about five years. Individual landowners or small-scale local cooperatives will hardly have a chance to raise the necessary equity contributions for projects of this size. This is even more so the case in a state like Brandenburg, where the population earns 20,000 Euros less than the German average per year, and the accumulation and inheritance of financial capital is well below that of their West German cousins, due to the 40-year history as a socialist country. With financial wealth being unequally distributed – some people owning large sums, looking for ways to place their funds and generate revenue, and the rest lacking substantial savings – and banks hardly offering interest, the revenues of up to 6 or 7% in the renewable energy sector sound very promising.

Project developers flock to places like sparsely populated Brandenburg to find owners and communities willing to rent or sell their lands for energy production. Parts of the profits will go to the individual owners: rents for one wind turbine are as high as 80,000 Euros per year for the half hectare needed, or 3,000 Euros per year for the hectare needed for solar panels. Furthermore, the communities within 3 kilometres of a wind park receive 10,000 Euros per newly erected wind turbine per year, and an additional (optional) 0.2 cent per produced kWh. Tax income, on the other hand, is often less than promised and comes comparatively late, as companies amortise their assets over years. The major share of income, however, remains with the company planning, installing, and operating the renewable energy project. It is also not uncommon to sell wind and solar parks after installation, making it even harder for communities to guarantee an income from local taxes, for instance. The returns are indeed substantial: a wind park with 20 large turbines, like the one planned in Welsickendorf, would generate a turnover of 13 million Euros per year, leaving after subtracting all costs, like service, revenue, and rents, a large single-digit million Euro sum. The wind park would be amortised after 12 years, leaving this sum for the next eight years as profit for the company.

At the same time, the rents landowners receive will be, with about 50–60,000 Euros for one plot, significantly higher than the average yearly income in Brandenburg, making it easy to understand that those owning land in a designated area are often more than willing to rent out their plots – especially if conflicts with other forms of land use, for example, farming, do not arise. In this respect, renewable energy companies are able to leverage regional disadvantages. However, as not everyone owns land or is 'chosen' for the wind turbine locations, there is another dimension of distributional injustice, which sits within the village or community. Everyone living there will be bearing the potential burden, but only some will profit financially. Prokon's 20:80 model is designed to address this and offers a way forward as it includes everyone owning land in the designated area and not only those whose small plots are chosen to site turbines. But as noted, it is not an ultimate solution, allowing everyone affected (including those who own no land at all) to profit from renewable energy production. The Brandenburg's 2019 wind energy levies act also works towards levelling distributional injustices in this regard, as it involves an even larger number of people financially, as do the 0.2 cent per kWh for the affected communities.

Finally, in Brandenburg, conflicts over the distribution of revenues and profits from wind energy projects, together with questions of planning and ownership, were more significant than those of renewable energy as a sustainable employment sector for local people or regional development, as we have been observing in South Australia and Karnataka. It was never an issue that arose in our conversations, and in our 2021 questionnaire, workforce issues and regional development were the two aspects that people saw least affected by wind energy production, neither in a positive nor negative way.

Solar Park Planning – 'I don't Like Cooperatives, but for the Farmers It Would Be a Very Good Option'

Solar energy production in Brandenburg has a very different history and logic. While greenfield solar parks, like wind farms, offer hardly any local job prospects, they – in contrast to wind turbines – tend to be directly in conflict with farming. Solar is currently shifting from individual households and industrial roof-top solar, and regulated installations along railway tracks and autobahns, to unregulated greenfield mega projects. Something like a solar boom is underway. Again, examples from the region of Teltow-Fläming are a case in point.

Jens and his son's company farms about 800 hectares of farmland in Teltow-Fläming. They concentrate on conventional crop farming but diversify their business with reloading and stocking wheat for other farming companies and have experimented with ginseng and cannabis farming. Jens inherited about 75 hectares from his parents and bought and rented additional plots after reunification in 1990. Producing wheat, sugar beet, rapeseed, maize, rye, and other grains, they make a good living, but their earnings depend on weather conditions, world market prices, and European agricultural subsidies. In late 2021, the mayor invited Jens together with all other farmers in the community to a meeting on greenfield solar. The idea was to get an overview of the farmers' perspective on greenfield solar installations, which had become a focus for energy investors across Brandenburg.

The mayor and the local council previously had decided in a council meeting that they would not be willing to grant any building permit for solar photovoltaic panels on agricultural land until a national regulation was in place. However, Jens and other local farmers – ranging from comparatively small businesses farming 100 hectares or less to large ones tilling 1,700 hectares and more – rejected such a ban. They were, on the contrary, open to renting out (parts of) the land they owned to solar energy companies. Their reasoning was economic, grounded in the calculus of capitalist accumulation: while they could realise a profit of about 200 Euros per hectare from conventional farming, solar companies would pay an average of 2,500 Euros rent per hectare for installing solar panels. Taking into consideration that the Brandenburg soil in some (but not all) areas is comparatively sandy and unproductive – a situation that might worsen with climate change – the idea of earning money without tilling operations and of diversifying your income sounded promising.

However, during and after the meeting, Jens raised an important point: as a farmer, it was possible to earn money from greenfield solar only if you are the owner of the land, not if you lease it from others. In Brandenburg, leasing is common, with about 73% of all agricultural companies leasing land. With roughly 1.3 million hectares for agricultural uses, almost 0.9 million hectares are leased by farmers (Statistiche Ämter 2023). Leases for small-scale individual owners, the church, or the community usually last 12 years. These leases are unlikely to be able to compete with solar investors offering to rent the land at up to 10 times the current rate. Furthermore, if the local municipal government grants the building licence for a PV park, the owner can immediately cancel the contract they have with the farmer. The conflict reflects the competing use values of land appropriated for solar power and agriculture. In a capitalist economy, without regulation or collective action, such conflicts are mediated through competition over land rents that are resolved according to which use is more profitable.

Farmers have opted to explore possibilities for collective action. 'There should be some form of farmers' co-operative', Jens explains, 'so the farmers don't lose out on this'. In general, he is not in favour of co-operatives, he says, but in this case some form of it might be a solution: all farmers and land owners needed to agree that they would inform each other when investors are approaching, so that contracts are only signed if the farmers who are habitually tilling the land get a share of the profits that the owners and investors make. If all agree that without this fair sharing there would be no land lease for solar PV at all in the area, then there could be no outrivalling each other, and local revenue for farming companies from solar PV would be maximised.

The community discussion about solar PV is situated in the specific local context of solar park planning, which is focused on 90 hectares of farmland that is tilled by three companies (one of them Jens's) and owned by about 30 individual landowners. A renewable energy planning company has reached agreements with most of the landowners for the installation of the solar park, which would generate about 90 MW_p. When the planning company introduced its park idea to the local council, it had offered share ownership to the local population: as the project planner informed us in an interview, 'the citizens of the community are given the opportunity to invest in the operating company'. However, to benefit from this, the citizens needed to invest their own capital. This excludes citizens who are not financially able to invest and compounds locational injustice: those residents who lack the capital to participate financially and benefit from the solar farm are still affected by its presence. Initially, the project planner also proposed to install solar panels on municipal buildings like the community school, framing this as a gift to the community. Yet, this seemed to have been taken off the table. According to the project planner, the municipality had decided to treat roof-top photovoltaics as a separate project, because the application processes were rather different. He explains,

> It's just two very different things. On the one hand, there are roofs that are already per se covered by the EEG [*Erneuerbare Energien Gesetz*, Renewable Energy Law], you no longer need a building permit to do that, it's simply regulated by the EEG.... My project is based on a development plan, because there is another type of use in the area. It's just this conversion of these areas, this process, it's completely different from the process for the solar systems on the roofs. (Interview Project Planner 2021)

This example sits within the larger framework of greenfield solar installations in Brandenburg, as well as the German government's ambitious goals concerning renewable energy production. The building of PV parks is not centrally controlled by the national government but technically falls to the concerned federal states; and these states, too, devolve responsibility and redirect decisions on photovoltaics that are not subject to the EEG – such as the one the planner introduced – to their municipalities. These hold the planning authority, negotiate conversions of land use, adjust legally binding land-use plans, and grant building permissions; in other words, this is another example of distributed energy governance.

While wind power installation nationwide has stalled at approximately 3% annual growth, open-field photovoltaic projects are gaining ground, especially in

Brandenburg. That is, for one thing, due to the topography of the state and its predominant land use forms. The state of Brandenburg is comparatively flat and includes large stretches of decommissioned military areas and former open-cut lignite mines that would (after being mined out) technically qualify for open-field solar parks. However, due to the additional costs of clearing former military areas or stabilising former mine sites, solar park developers target the 1.5 million hectares – 50% of the state – used for agricultural production.

As of now, the rush for open-field photovoltaic energy is hardly visible in the landscape. In Brandenburg, there is only one finished large-scale photovoltaic plant on agricultural land, in Werneuchen near Berlin. With 164 hectares, this is the largest in Germany to date. Yet, many others will follow: Overall, the number of accumulated inquiries in Brandenburg since 2019 amounts to at least 366 projects with a total area of more than 9,600 hectares. For 55 projects accounting for more than 2,800 hectares, installation procedures were underway in 2021, with some already having been approved (Keller, Huth and von Daniels 2021). In late 2021, project developers approached a municipality in Teltow-Fläming with the idea of installing a 1,000-hectare solar park. This would mean covering an area of 10 kilometres by 1 kilometre, or 1,400 soccer fields. In our interviews, some project developers used the image of a 'gold rush mood', but did so in order to differentiate themselves from other developers:

> [...] they [the municipalities] get a thousand applications from good companies, from bad companies, from some gold digger trying to sell something. And I have to set myself apart, I just have to tell the community: Hey, you shouldn't experience the same thing as you did with the large wind farms, where a lot was promised and nothing happened, but more has to happen now (Interview Project Planner 2021)

The topography of the state together with the cheap sell-off of public lands in the last three decades makes Brandenburg particularly attractive for large solar park installations. Yet another reason for the rush for open-field solar is the state's land use history. Brandenburg is an energy state with a long history of lignite mining and a shorter history of renewable energy production. In consequence, grid connections are available, making new solar park installation more economic. Additionally, due to the history of forced collectivisation in the German Democratic Republic, comparatively large agricultural production cooperatives had emerged after 1990. The sheer size of the areas covered by the cooperatives makes the agricultural land in the East interesting for solar investors. Developers need large tracts of land – typically more than 50 hectares – to make a return, to spread planning and installation costs across a large area. Due to the drop in the cost of PV modules and the lack of regulations for unsubsidised solar parks, large ones have become very attractive for investors, as outlined by one head of an online PV platform, Milk the Sun, which connects affluent capital owners with project developers:

> What we're seeing here is a break-out into the pure market, which of course means that I have to find the right location, there are no limits on size, at least no legal limits, and that brings a whole new set of challenges with it. That's exactly what we're working on now with new project developers; we look for investors, and for partners who are able to take on these projects, which are getting bigger all the time, and make them happen (Interview Head of Milk the Sun 2021).

In order to be able to build PV panels on agricultural land, land use has to be officially converted by the respective local government. For such conversions, the relatively poor soil quality of large areas in Brandenburg is a reoccurring factor. Around 80% of Brandenburg's agricultural area has been designated as disadvantaged since 2018, due to natural conditions with low-yield and often sandy soils as well as regional climatic factors (MLUK 2023). As reported by the German Weather Service, *Deutscher Wetterdienst*, an average temperature of 10.8°C (four-year average of 8.7°C) was measured in Brandenburg in 2020, while there was only around 499 l/m² of precipitation (with a four-year average of 557 l/m²) making it one of the warmest and driest federal states in Germany. In addition, frost events, early summer drought, and the increase in extreme weather events (e.g. heavy rain/hail) lead to a significantly increased production risk for agriculture (MLUK 2023).

In Germany, the productivity of agricultural land is evaluated numerically on a scale from 0 to 100. In Brandenburg, this productivity scale averages less than 35 per hectare. Thus, a farmer in Teltow-Fläming with a soil quality of 32 has considerably lower cropping capacity from their fields than for instance a farmer in the Magdeburger Börde in the state of Saxony-Anhalt, which has a soil quality of 100. Low-productivity areas are officially considered disadvantaged according to European Union regulations. To ensure that such agricultural areas do not fall fallow, but continue to be cultivated, the state can grant a subsidy, and Brandenburg does with the so-called compensatory allowance (*Ausgleichszulage*). Farmers receive 25 Euros per hectare in disadvantaged areas. However, even considering the government's subsidies, farming cannot financially compete with the return from solar PV on these fields.

The Case's Energy Conflict Forms

The pressure from PV investors increases as funds are collected by prospective investment companies anticipating large profit margins. In this context, landowners can expect to gain a substantial increase in rental income: the solar companies offer a lease of between 2,000 and 3,000 euros per hectare for the next 30 or 40 years. Again for comparison, a solar park with 50 hectares could generate annual earnings of 125,000 Euros for the landowner, as compared with

18,000 Euros when renting it out (or about twice that much when farming the land themselves). Depending on the concentration of land holdings, only a handful of individuals would receive this considerable increase in income with PV installations. Accordingly, PV threatens to displace existing leaseholders, make redundant existing farm workers, and deliver windfall, unearned rents to landholders. It creates potential for conflict within and among affected communities living close to the envisioned projects that can undermine legitimacy to the extent that this threatens the viability of the development. The specific distributional conflicts reflect the characteristics of solar appropriation, where it remains difficult (if not impossible) for large-scale solar power to coexist with agriculture.

With this, solar power creates a specific regime of accumulation in terms of who reaps the financial rewards and who is displaced from their livelihood or forced to adapt. The parallels with the Karnataka experience are clear, in terms of displacement of less powerful players, and also in terms of prefiguring the necessity for more participative models for shared benefit. The emerging model of solar appropriation and its displacement logic make it possible to argue for alternative social ownership models where shared benefit is foundational rather than simply a gift dependent on goodwill or political favour. As evidenced in distributed renewables, benefit-sharing is socially contagious and could be used to guide renewable policy rather than be seen cynically as an enabler for private interests.

Given the role of private generating companies, benefit-sharing is more commonly deployed to gain local consent and insulate renewables proponents from opposition. Compensatory inducements have become an important aspect of gaining an edge against competing projects; as one developer put it:

> You can find this potential for envy relatively quickly in these municipal council meetings, that's easy, of course, nobody has anything against it, everyone wants photovoltaics, but now one farmer alone has a subsidy who earns money with it and everyone else doesn't. And I think that's the great tension where the community also says: Yes, why should we do that? Why should we do something now, that concerns us all. There are always people who are fundamentally against everything, you have to deal with them, that's also legitimate. Or maybe they have real reasons why they are dealing with it in such a planning process. It's also about several months or years and if I don't offer the community anything, as I described earlier, maybe a little more, because at the moment they are all, I believe, you can call any mayor, bombarded with inquiries. (Head of Milk the Sun, Interview 2021)

Here it seems as if project developers for solar PV have learned from the mistakes that were made with the implementation of wind energy productions in Brandenburg. As stated earlier in contrast to the wind farms, 'where a lot was promised and nothing happened', some developers see it as their duty to come up with solutions in order to come to an agreement with the community, though it should be noted that, similarly to the Prokon wind project described earlier, this does not solely happen out of goodwill but as a strategy to avoid resistance, or to steal a march on competitors. Additionally, the idea of a farmers' cooperative that Jens proposed sounds like a promising model to reduce distributional conflicts and appears as yet another aspect of learning from the experience with wind parks as to how the residents can gain greater leverage in negotiations by acting collectively rather than alone.

As was stated in an interview earlier, 'nobody has anything against' PV, and our research does confirm that there does not seem to be any fundamental conflict over the use or rejection of photovoltaic as a source of energy. However, the more extensive, exclusive appropriation of land necessary for solar energy compared to wind can create acute contestation between actors with interests in other economic activities, principally agriculture. What appears first as an economic conflict over land use, rather than the aesthetic concerns of residents with wind farms, is nonetheless deeply imbricated with rural identities as farming communities. And, in terms of changed land use with photovoltaic energy on agricultural land, locational conflicts do become visible:

> When I want to build on agricultural land everyone says: Oh, agriculture, for god's sake, no, we need agriculture. Nature conservation says: Hey, nothing better can happen than when such monoculture landscapes disappear and the soil can recover there and a bit of biodiversity can arise again, when they are no longer fertilized. That's always the balancing act that one has to do. (Interview Head of Milk the Sun 2021)

Reflecting this position, the NABU, Germany's largest NGO for nature and environmental protection, spoke in favour of open-field solar in their position paper on solar parks 2022 (NABU 2022). In their position paper, the NABU pointed to the fact that if it is emphasised that solar parks consume agricultural land, it should also be taken into account that 60% of the agricultural land in Germany is currently used for fodder cultivation and a further 14% for 'energy crops' with only 22% used directly for food production (NABU 2022: 8). As NABU also points out, there 'is a fundamental need for change here in favour of more climate and nature protection in the agricultural sector, [and] solar parks can make a contribution to this' (NABU 2022: 8). In doing so, they are not rejecting outright the appropriation of land for economic production but analysing its social purpose and ecological implications.

The question of how arable the land is in terms of what is actually farmed is a reoccurring theme in the discussions over open-field solar. However, photovoltaic project developers see both aspects as favouring their interests. As noted, large areas of Brandenburg do not have arable land of the same quality as in other parts of Germany, while many of the farmed products are actually used as fuel. Currently, 17–20% of Brandenburg's agricultural land is used for energy crops, especially maize for biogas plants, producing monocultures with little ecological value (Land Brandenburg 2010). However, photovoltaics are 30–40 times more efficient per unit area than biomass, meaning that in future, with electrification and a switch to PV generation, only a fraction of the energy crops will be needed. Project planners often use these sorts of projections in community council meetings to convince residents and their representatives that change is inevitable. As one project planner put it:

> It is very exciting to do a bit of education, to say: Folks, there are also agricultural areas that are not so incredibly valuable, that have low *Bodenpunkte* (official land value) and in case of doubt, corn or rapeseed for biodiesel is now being grown. I am convinced that it will never end up on the table, but that it will go into the tank and be burned. Agriculture that has nothing to do with nutrition, nothing with food for animals or plants, but rather fuel is produced. I have to say, from my point of view it doesn't really get any more stupid, but you have to look at how high the energy density is, so to speak, to bring back such areas to say: Okay, if I actually don't need the land for food production, I have to say that I can use parts of it to build PV on it, as we do on a large scale in Brandenburg. (Interview Head of Milk the Sun 2021)

However, understanding socio-ecological systems as an assemblage opens up ways to see regional solar park installation as more than soil productivity numbers or fodder versus food cultivation. Solar park installation is part of, and also influences, the entanglements between the historic development of village life and agriculture, before and after GDR times, the resulting investment options and potential, cooperation and competition among farmers and neighbours, and, again, rural development and community values. As Martin, the head of a spiritual retreat place in Teltow-Fläming, explains in light of a planned solar park:

> In light of the current solar mega projects underway in Brandenburg, the existing laws to regulate greenfield solar PV are toothless tigers. The Federal Environment Ministry has stated in the Renewable Energy Sources Act that the expansion of open space systems should only take place to a limited extent, and that the expansion of roof systems should be strengthened. The ministry explains that for environmental and nature conservation reasons, the expansion of PV from the rooftops to open spaces should be avoided. Reflecting this, the public subsidy for open-field PV is limited to installations below 20 megawatts. The separate tender segments for roof-top photovoltaics and open-space photovoltaics was designed to assist in avoiding open-field PV installations. However, projects that could be economically sustainable without EEG funding are not limited by this law.

Instead of clear legal requirements, there are several non-binding recommendations hinting at possible procedural conflicts over open-field photovoltaic projects. The regional planning community Uckermark-Barnim was the first to publish procedures. Later, the Brandenburg Greens also published a guide and some communities have set their own criteria, limiting photovoltaic projects to arable and grassland areas of low quality, that is, with a maximum of 28 *Bodenpunkte*.

Such guidelines, which should arguably also be expected in legally binding form from the Brandenburg state government or the German national government, point not only to soil quality, financial benefits, and productivity of farmland or energy generation but also reflect latent identity conflicts. While Brandenburg calls itself an energy state, its landscape and hence a large aspect of visually perceivable and locally embodied identity is dominated by farming, even though farming is done by only 2% of the population but covers 50% of the surface. Hence, not only for those who still work in or in relation to farming in Brandenburg, but also for those living in Brandenburg's villages and towns, the conversion of agricultural landscape to a solar park landscape entails an identity conflict. As the farmer Jens said:

> There was always this debate 'from agricultural farmer to energy farmer'. But nothing happened. I mean, who built his own wind turbine? I mean maybe providing land for lease, but that was it. And now there is this run for land for solar. And I'm saying, also to people approaching me: You take away my basis for production! (Interview Jens 2021)

The Brandenburg Ministry of Agriculture, Environment and Climate Protection also acknowledges the impact of green-field solar parks on people's identities. In its 2021 provisional guidelines, it recognises that if the new forms of land use develop in a comparatively short time frame, there will be an increased need for preventive and orderly planning and control. Furthermore, any such change will require 'strong public acceptance for the related changes in people's living environment' (MLUK 2021: 2). The guidelines recommend a careful calculation of positive and negative aspects, with a strong preference for the installation of solar parks on former military or industrial lands; they also indicate that communities should have the means to prevent or break up proposed parks that are too large (similar to guidelines of Uckermark-Barnim; MLUK 2021: 2–10).

It remains an open question if, over time, residents of regions such as Teltow-Fläming will come to see the changing appearance of the landscape, and the increasing presence of solar and wind energy, in a positive way, or whether it will require too much of a change in their sense of local identity, especially in the context of the other energy conflicts we have outlined. Would camouflaging large solar parks with hedges, for instance, be one way of cushioning them from identity conflicts? Will ownership models, and future preventive and ordering regulations, minimise distributive and procedural conflicts, and allow Brandenburg's citizens to benefit from and positively take up the idea of the state as an energy state in the field of green-field photovoltaics?

Wind and Solar – Continuous and/or Resolved Energy Conflicts

As we have seen, the socio-ecological appropriation of 'nature's free gifts' for renewable energy in Brandenburg has opened up a number of different fields of contestation. We have explored the way this process of contestation is experienced, what interests are at stake, and what kinds of actors are drawn into the process. We have also highlighted the ways in which the dynamics of appropriation, accumulation, and regulation, which we identify in Chapter 1, underlie and drive these processes of contestation intertwined with the types of energy conflicts identified by Becker and Naumann.

Each of these dimensions of contestation or conflict has the potential to undermine – or strengthen – the social legitimacy of a particular project and of the *Energiewende* more broadly. While the state, in the form of federal and state governments and their agencies, has sought to foster the legitimacy of the German energy transition as a *national* project, especially over the last decade, we argue that legitimacy is also created from below, through the experiences of citizens confronted with the development of renewable energy in their everyday lives. Moreover, it is at the local level – and primarily through the regional planning authorities – that the energy transition is regulated and governed, and the 'social making of arguments' occurs, and with it, the social contest over legitimacy.

The conflicts we identify in our field study locations, and in Brandenburg more broadly, play out over three broad phases. In the first phase, which we can label 'Wind 1.0', a wind rush occurs, beginning slowly at first, but then gathering pace, and is characterised by a failure to foresee the consequences of planning processes and decisions on the ground. The arrival of wind energy installations in the energy landscape divides and stratifies affected communities and alienates some local residents. As a result, renewable energy becomes politicised and the social legitimacy of the energy transition is called into question. In the second phase, which we describe as 'strategic engagement', project developers and local and regional authorities attempt to bolster the legitimacy of renewable development – in this phase, still primarily wind energy – by addressing some of the grievances of communities. At the same time, community members begin to organise and mobilise. In the third phase, a solar rush begins, which we label 'Wind 2.0', the term used by Frank Deutschmann, mayor of Heideblick. This third phase generates some of the same conflicts as the wind rush, but also some new strategies and responses from solar developers and affected communities, suggesting that both have taken some lessons from the experience of the previous two phases.

As we note in Chapter 1, there is no renewable energy without access to land; nature's free gifts cannot be appropriated and capitalised without land on which to site wind and solar installations. The renewable technology-ecology nexus takes concrete form in energy landscapes which, in Brandenburg at least, are predominantly rural in nature,[1] but which are also inscribed with complex social and

[1] While this may seem obvious, in Australia energy landscapes are also urban; cf. the unique role of rooftop solar, primarily in Australian cities, in Australian renewable energy production.

economic relations, shaped, in particular, by the distinctive history of Brandenburg during and after the socialist period from 1945 to 1989. As we noted earlier, the communities and community members in our study sometimes mobilise concepts of nature and rural identity strategically as a way of resisting renewable energy development; ecology is mobilised against technology. Our findings resonate strongly with one of the key recommendations made by the expert commission 'Energy of the Future', referred to in Chapter 1, namely that strengthening the social acceptance of renewable energy in the places where it is produced – in other words, in the countryside – will be decisive for achieving the rapid expansion of renewables necessary to meet Germany's climate and energy goals. That recommendation was made specifically in relation to wind energy, but as our study shows, it applies equally to the more recent expansion of large-scale solar energy. Social legitimacy is not only required to get individual projects over the line, but to maintain ongoing public support for what is a multi-decadal energy transition.

The regulation of renewable energy has been a key focus of contestation in Brandenburg. The state of Brandenburg has actively promoted the expansion of renewables, until recently primarily wind energy, and promoted itself as an energy state. But at the local level, where that energy is produced, the forms of regulation and planning (described earlier in the section 'Wind Park Planning I') implemented by state agencies have often had the effect of intensifying rather than reducing contestation, especially in the form of locational and procedural conflicts around wind development.

Flows of investment capital into the energy landscapes in our case studies have clearly created or exacerbated distributional conflicts. Rather than democratising access to the utilisation of nature's free gifts, they have tended to benefit particular landholders who can lease their land to project developers, while others living in close proximity to wind or solar projects receive lesser or no benefits at all. As we argue, the reasons for this lie in the amendments and partial insufficiency of the Renewable Energy Law and the concentration of wind parks, as well as in the distribution of financial capital and expertise. Unequal access to capital is a key issue here; local communities which might want to develop community-owned renewable energy projects generally do not have the same ability to raise capital as corporates. More broadly, the 'uneven development' of western and eastern Germany has tended to mean that communities such as those in Teltow-Fläming are effectively on the periphery of capitalist development and of energy capital, at least in terms of participating and benefiting financially. The expansion of renewable energy has therefore tended to deepen, rather than ameliorate, existing historical and geographical patterns of uneven development in the region.

Creation of local employment and the flow of excess energy to downstream industry are not significant themes for our interlocutors. Comparing our ethnographic studies of solar and wind development in Brandenburg, we can see that conflicts around green-field solar installations display some of the same features as earlier conflicts over wind development. However, the companies have learned from previous experiences, and tend to involve affected residents earlier in the planning process. In the past, the lack of opportunities for participation has been one important factor in dissatisfaction with the development of wind energy and has been a reason for citizens to contest new wind parks in Brandenburg. But project developers in both the wind and solar industries are now trying to pre-empt these procedural conflicts by informing and involving residents earlier.

Therefore, we can see these procedural conflicts as also having a productive dimension, in that they push regulatory processes for renewable energy development in a more democratically responsive direction. The evolution of a distributed system of energy governance, mirroring the distributed nature of wind energy generation, has created new opportunities for mobilising innovation. To the questions we posed in the introduction to this chapter – whether renewable energy project developers, planning authorities, and local communities have learned lessons from the expansion of wind energy in Brandenburg, and whether our case studies show that new forms of development and social engagement with renewables are possible – we answer a qualified yes.

Here it is worth reflecting for a moment on how these case studies fit within the historical development of renewable energy in Eastern Germany, and the forms of energy governance which have evolved around it. The agreement reached in the village of Feldheim between the project developer Energiequelle and local stakeholders – villagers, local government, and the farmers' co-operative – is often seen as a successful model of renewable energy development, which avoided the kinds of conflicts we have outlined earlier and strengthened rather than weakened the social legitimacy of the energy transition at a local level.

But the 'Feldheim model' is not necessarily one which could be duplicated now. The wind park there has evolved over nearly 30 years, from an initial installation of four small turbines to 60 much larger ones today, and its evolution relied on a specific set of conditions: the subsidies available at the time, including funds from the Brandenburg state government and the EU, the lower capital cost of the turbines themselves, willingness on the part of the developer to cooperate and co-own with the local community, and the positive attitude of the community itself. Today, the turbines which have become the industry standard are themselves much larger and carry much larger investment costs; feed-in tariffs have been replaced by reverse auctions, the planning system has become much more complex, and the regulation of the electricity grid has changed. All of these factors mean that simply attempting to reproduce the Feldheim model in the 2020s would be nigh on impossible.

In one sense, we could see Feldheim as an example of what Geels (2004) describes as the transition from niche to regime in socio-technical systems; in the 30-odd years since the first turbines were installed there, wind power has advanced from a niche technology, a 'locus of radical innovation' around which the rules have not crystallised and which receives some forms of protection (Geels 2004: 912–913) to an integral component of a new energy system, in which various forms of renewable energy have become part of the dominant 'regime'. In Geels' terms, renewable energy has 'broken through' into the 'mass market' which is the German and European energy market; 'a prolonged period of uncertainty, experimentation, and co-existence of multiple technical options' is underway in the energy system, but a new socio-technical regime has yet to stabilise (Geels 2004: 916).

It could be argued that the shift from niche to regime, the growing involvement of finance capital, and the pressure for efficiencies of scale (for example, in the form of much larger turbines) inevitably foreclose the kinds of social innovation which Feldheim represents. Certainly, the law passed by the Brandenburg state government in 2019 with the specific aim of 'increasing the acceptance of wind energy installations', and requiring developers to pay a fee of 10,000 Euros per wind turbine to local councils within 3 kilometres of an installation, helps a little to change the dominant socio-ecological relations of renewable energy. On the other hand, examples such as the municipal development contract developed by Frank Deutschmann in Heideblick or the willingness of farmers such as Jens to consider forming a farmers' co-operative suggest that, even under the present conditions in which renewables have become part of a gradually stabilising energy regime, some forms of social innovation and experimentation are still possible.

Open-field photovoltaic is a relative newcomer to Teltow-Fläming and other parts of Brandenburg, but wind energy has been a growing presence in the landscape for 20 years, and increasingly a focus for contestation as numbers, height, and concentration increased. Previous experiences with wind development – the way in which it was conducted by project developers, the extent to which local communities had a say in the development process, and who benefitted from it – have shaped attitudes to further development of the energyscape, not just in this region, but more broadly throughout Brandenburg.

Community ownership, or the kinds of benefit-sharing arrangements proposed by Prokon in the second case study, may go part of the way towards addressing distributional conflicts as they are experienced at the local level and increasing social acceptance of new projects. But they do not address the more fundamental issues of structural inequality and access to capital which we have outlined earlier, nor do they challenge the dynamics of appropriation, accumulation, and rivalry, which, as we have shown, tend to exacerbate existing inequalities and erode social legitimacy. The capital-nature 'frontier' of wind energy has become well-established in Brandenburg over the past three decades, and it appears that a new frontier of solar energy development is rapidly opening up, even during the course of our research and fieldwork. But our research suggests that unless the proponents of 'Wind 2.0' are prepared to learn from the experiences of 'Wind 1.0', they will encounter similar conflicts and similar erosion of the social legitimacy of this new generation of renewable energy.

Community ownership also does not address the issue of 'urban-rural exploitation', which we touched on earlier. The spatial fix which has been created for renewable energy in Brandenburg means that, while wind and solar farms are situated primarily in rural areas, and the burdens, both material and perceived, are born by people in the countryside, the majority of that energy is consumed by urban populations, in particular the residents of Berlin and Potsdam. This, combined with the fact that the cost of connecting installations to the grid is borne primarily by people in the countryside, in the form of higher electricity prices, creates resentment and can potentially undermine the social legitimacy of the energy transition in the very local arenas, which are crucial to its further development.

In conclusion, we have sought to address problems of legitimacy in renewable energy, focusing on the underlying effects of renewable appropriation, accumulation, and regulation. Our ethnographic lens has yielded insights into relations and refractions across the field of legitimation and agency, as an assemblage of possibility, or 'promise'. With this we have gained a more nuanced comprehension of prevailing energy conflicts, and also of their generative potential. The resulting narratives of 'renewables learning', across phases in its development, point to new participative configurations of future 'renewabilities'. By acknowledging the emergence, heterogeneity, and decentralised nature of social dynamics, we transcend conventional policy recommendations centred on public 'acceptance' of a corporate-centred status quo. Instead, we illuminate the intricate interplay across technology, human and non-human actors, landscapes, and environments that define energy regions.

Overall, our approach has sought to ground the legitimacy problems of renewable energy in the underlying socio-ecological antagonisms of energy transition. These stem, as we have argued, from the contradictions of appropriation, accumulation, and regulation. Understood dialectically, these generate new agendas and possibilities for renewable transformation, unveiling new trajectories for renewable energy development. Distributional conflicts, particularly evident in the context of wind energy, encompass both financial disparities among stakeholders and the spatial distribution of turbines within regions, adding layers of complexity to the assemblage. Furthermore, this framework elucidates the transformation of rural landscapes and communities in response to energy projects, highlighting that the energy transition extends beyond the adoption of sustainable technology, encompassing historical, economic, and social dynamics. Ultimately, our approach underscores the necessity for holistic, context-specific analyses, arguing the need for more ambitious holistic and transformative approaches to meet the multifaceted challenges of energy transition implementation.

The sheer scale of new wind and solar installations required to accelerate the energy transition and meet Brandenburg and Germany's urgent emissions reduction goals over the next decade *appears* to necessitate a particular spatial, temporal, and social fix: big projects financed by big capital. Whether other solutions, involving more distributed and democratic models of generation, ownership, and governance, could offer an alternative is an open question, which we return to in Chapter 6.

References

Becker, S. & Naumann, M. (2018). Energiekonflikte erkennen und nutzen (Identifying and Using Energy Conflicts). In O. Kühne & F. Weber, eds., *Bausteine der Energiewende (Building-blocks of the Energy Transition)*, Wiesbaden: Springer VS, pp. 509–522.

Boyer, D. & Howe, C. (2019). *Energopolitics. Wind and Power in the Anthropocene*, Durham: Duke University Press.

Brandenburg State Parliament (2019). *Act on the Payment of a Special Levy to Municipalities in the Vicinity of Wind Turbines (Wind Energy Taxation Act)*, Potsdam: GoB.

Bues, A. (2020). *Social Movements against Wind Power in Canada and Germany*, Oxford: Routledge.

Chezel, E. & Labussière, O. (2018). Energy landscape as a polity. Wind power practices in Northern Friesland (Germany). *Landscape Research*, 43(4), 503–516.

Cronon, W. (1995). *Uncommon Ground. Rethinking the Human Place in Nature.* New York: W.W. Norton & Company.

DW (2015). 'Investors rescue wind developer Prokon', Deutsche Welle (National Public Broadcaster), 2 July 2015.

Deutsche WindGuard (2023). *Status des Windenergieausbaus an Land in Deutschland Jahr 2022*, Varel: DWG.

Federal Parliament (2022). *Law to Increase and Accelerate the Expansion of Wind Turbines on Land*, Berlin: Bundestag.

Futrell, R. (2003). Framing processes, cognitive liberation and NIMBY protest in the US chemical-weapons disposals conflict. *Sociological Inquiry*, 73(3), 359–386.

Gailing, L. & Moss, T. (2016). *Conceptualizing Germany's Energy Transition: Institutions, Materiality, Power, Space*, London: Palgrave.

Gailing, L., Bues, A., Kern, K. & Röhring, A. (2019). Socio-spatial dimensions in energy transitions: applying the TPSN framework to case studies in Germany. *Economy and Space*, 52(6), 1112–1130.

Geels, F. (2004). From sectoral systems of innovation to socio-technical systems insights about dynamics and change from sociology and institutional theory. *Research Policy*, 33(6), 897–920.

Geertz, C. (1998). Deep hanging out. The New York Review, 22 October 1998.

Heideblick Municipality (2018). 'Resolution to conclude an urban development contract: Wind energy use area W 13 Falkenberg/Pitschen-Pickel, with the company Energiequelle GmbH', ALLRISNET, 18 June 2018.

Helfer, F. (2022). Using assemblage theory in political anthropology. Assemblages as claims on scientists and readership. *Hamburger Journal für Kulturanthropologie*, 14, 33–47.

Hübner, G. (2012). Die Akzeptanz Erneuerbarer Energien. In F. Ekardt, B. Hennig & H. Unnerstall, eds., *Erneuerbare Energien – Ambivalenzen, Governance, Rechtsfragen*, Marburg: Metropolis, pp. 105–127.

Jungjohann, A. & Morris, C. (2016). *Energy Democracy: Germany's Energiewende to Renewables*, Wiesbaden: Springer.

Keller, G., Huth, K. & von Daniels, J. (2021). 'Glowing landscapes: This is why solar investors endanger nature and agriculture in Brandenburg', Correctiv, 13 October 2021.

Kern, F. & Rogge, K. (2016). The pace of governed energy transitions: agency, international dynamics and the global Paris agreement accelerating decarbonisation processes?. *Energy Research and Social Science*, 22, 13–17.

Küster, H. (2010). *Geschichte der Landschaft in Mitteleuropa: Von der Eiszeit bis zur Gegenwart*, Munich: Beck.

Land Brandenburg (2010). *Biomassestrategie des Landes Brandenburg (Biomass Strategy for the State of Brandenburg)*, Potsdam: Ministry of Environment, Health and Consumer Protection.

Leibenath, M., Wirth, P. & Lintz, G. (2016). Just a talking shop? Informal participatory spatial planning for implementing state wind energy targets in Germany. *Utility Policy*, 41, 206–213.

Löschel, A., Grimm, V., Lenz, B. & Staiß, F. (2021). *Stellungnahme zum achten Monitoring-Bericht der Bundesregierung für die Berichtsjahre 2018 und 2019*, Berlin: Expert Commission for the Energy of the Future Monitoring Process.

Marcus, G. & Saka, E. (2006). Assemblage. *Theory, Culture & Society*, 23(2), 101–106.

MLUK (2021). *Vorläufige Handlungsempfehlung des MLUK zur Unterstützung kommunaler Entscheidungen für großflächige Photovoltaik-Freiflächensolaranlagen*, Potsdam: Ministry of Agriculture, Environment and Climate Protection of the State of Brandenburg.

MLUK (2023). *Agrarbericht: Natürliche Bedingungen*, Potsdam: Ministry of Agriculture, Environment and Climate Protection of the State of Brandenburg.

Müller, K. & Morton, T. (2018). At the German coalface: interdisciplinary collaboration between anthropology and journalism. *Energy Research & Social Science*, 45, 134–143.

Müller, K. & Morton, T. (2021a). *Brandenburg-Report: Zum Stand der Windenergieerzeugung in Brandenburg, Centre for Interdisciplinary Regional Studies*, Halle: Martin Luther University Halle-Wittenberg.

Müller, K. & Morton, T. (2021b). The space, the time, and the money: wind energy politics in East Germany. *Environmental Innovation & Societal Transitions*, 40, 62–72.

Müller, K. & Pampus, M. (2023). The solar rush: invisible land grabbing in East Germany. *International Journal of Sustainable Energy*, 42(1), 1264–1277.

MWAE (2023). *Renewable Energy*, Potsdam: Ministry of Environment, Health and Consumer Protection.

NABU (2022). *Position Paper on Solar Energy*, Berlin: The Nature and Biodiversity Conservation Union.

New Energy Forum Feldheim (n.d.). 'The energy supply of the energy self-sufficient village of Feldheim via private local heating and electricity networks: success through cooperation', Feldheim: NEFF, n.d.

Renn, O. (2011). The social amplification/attenuation of risk framework: application to climate change. *WIREs Climate Change*, 2(2), 154–169.

Reusswig, F., Braun, F., Hegera, I., Ludewig, T., Eichenauer, E. & Lassa, W. (2016). Against the wind: local opposition to the German Energiewende. *Utilities Policy*, 41, 214–227.

Schöbel, S., Hübner, G., Leschinger, V., Barthel, P. et al. (2022). *Aktive Bürgerexpert: innen in Klimaschutz und Energiewende*, Freising: Forschungsbericht.

Statistiche Ämter (2023). *Farms and Their Agricultural Operations: Used Area According to Ownership and Lease Relationships*, Potsdam: Statistiche Ämter des Bundes und Länder.

Windmesse (2022). 'Brandenburg: The moratoria for wind turbines was terminated by the Joint State Planning Department Berlin-Brandenburg for all affected planning regions', Windmesse, 2 December 2022.

Wolf, I. (2020). *Soziales Nachhaltigkeitsbarometer der Energiewende 2019: Kernaussagen und Zusammenfassung der wesentlichen Ergebnisse*, Potsdam: IASS.

5
South Australia

Louise comes from a multi-generational Port Augusta family and has worked in renewable energy and regional development projects. She speaks to us of the mocking references to her hometown that she experienced when living away for study, stigma born from the town's reputation as a 'dirty, industrial, working-class town'. She is committed to making the town a better place for the future of her children:

> … my family won't leave Port Augusta, and I want to be with my family, so I need to make Port Augusta better so that my children want to live here. I've always been embarrassed to say I'm from Port Augusta. Growing up, going to uni or anything like that, it was always one of those things, you were never proud of where you came from. I want my kids to be proud of where they come from, and I want them to have opportunity here. It's lovely that we're here…. My plan is to make the place the best it can possibly be. (Interview Louise November 2020)

Louise was elected to the Port Augusta City Council (PAC) in 2018. When we met with her in 2020, in the midst of Port Augusta's renewable energy construction period, she described renewable energy as an important part of rejuvenating the town's future. She wanted to see Port Augusta become the 'renewable capital of the world', stepping up from its present status as a drive-through town for 18,000 vehicles a day on the country's main north-south inland highway:

> From a personal perspective, and I suppose from a councillor perspective, I think: bring it on, keep it coming…. Renewable energy is one of the only industries really that we can start developing…. What have we got at the moment? We've got a rundown main street. We've got a corridor through town that's full of weeds. The whole place looks sad and rundown and why would someone stop? … I think renewable energy is only a good thing. Obviously, any development needs to be managed correctly. There's always going to be potential issues. (Interview Louise November 2020)

Louise reflects an optimism for the potential of renewable energy that we commonly found in the Upper Spencer Gulf (USG) region during our study, but so far it hasn't played out quite as it was hoped.

****

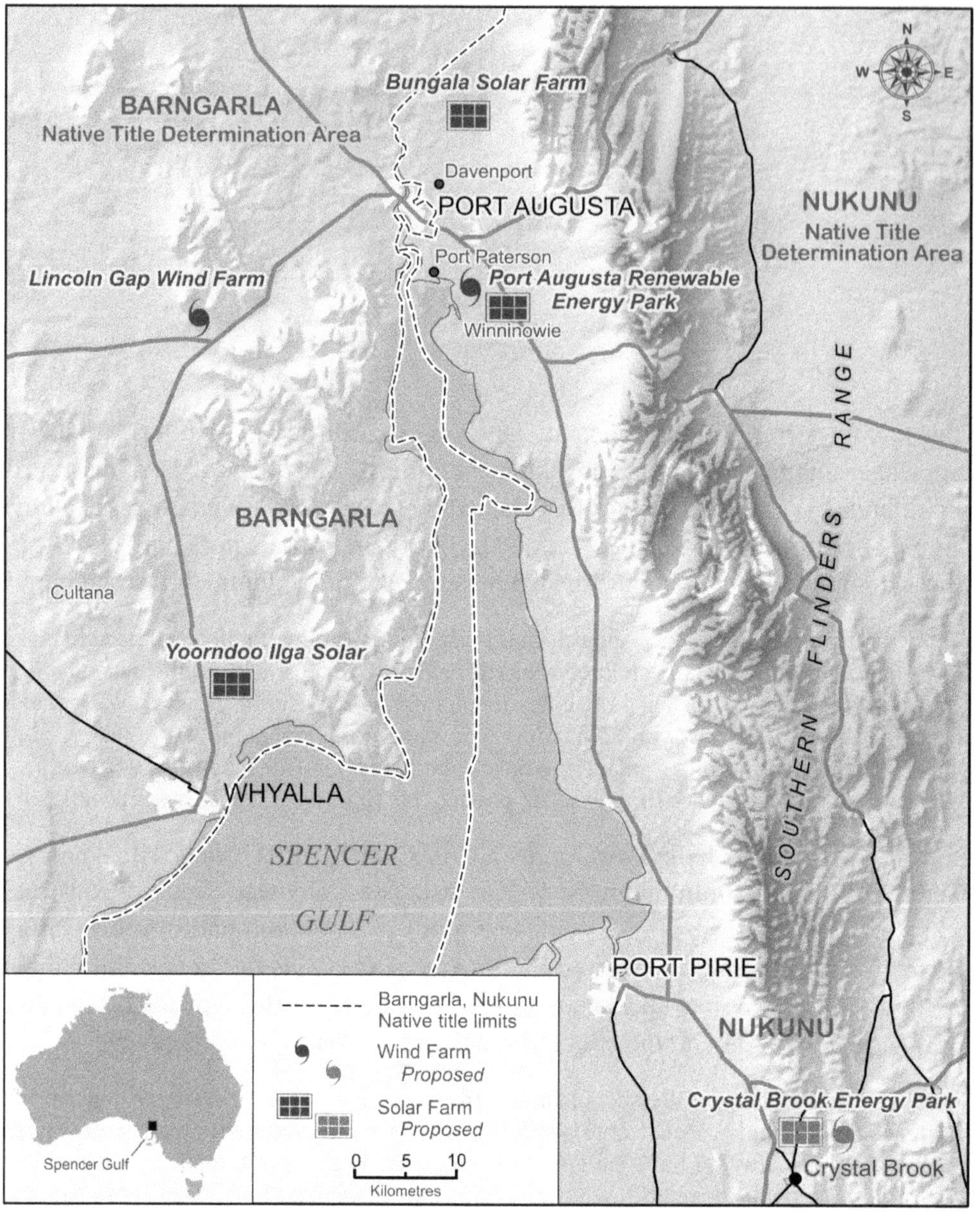

Map 5.1 Upper Spencer Gulf study area and selected renewable projects (Source: Olivier Rey-Lescure 2024). See Appendix A for the detailed list of large-scale renewables projects proposed for the USG.

This chapter examines the course of large-scale renewable energy projects in the USG of South Australia and the mixed outcomes for residents, Aboriginal Traditional Owners, businesses, local government, and others. Our field research extended over three years (2019–2022) through COVID delays. We witnessed a dynamic situation during that time frame, including the introduction of new technologies and the expanded scale of installations, two changes of state government, changes in planning regulations and procedures, successful native title determinations, and shifting patterns of corporate investment. Based on our ethnographic research and questions that apply across the three comparative regions in this book, we analyse the social legitimacy of decarbonisation as a socio-ecological transition in the USG. Key research sites appear in Map 5.1.

Upper Spencer Gulf Communities, Lands, and Livelihoods

Flanked by the Yorke Peninsula and Flinders Ranges to the east and the Eyre Peninsula to the West, the Spencer Gulf is the largest inlet on the coast of South Australia, the southern and central state of Australia. Locals commonly use the term USG to describe the broad geographical region located around the northern end of the Spencer Gulf (see Map 5.1). Like many non-metropolitan regions in South Australia, the USG population is relatively small, as 78% of the state's population resides in the capital, Adelaide (Australian Bureau of Statistics 2021). The USG population is concentrated in three small cities, each the centre of a local government area (LGA): Whyalla City Council (WCC), PAC, and Port Pirie Regional Council (PPRC).

Adjoining and economically linked to the USG is a huge arid area in the north of South Australia, comprising over 60% of the state. Most of this sparsely populated area is administered by a state statutory body, the Outback Communities Authority (OCA), and is commonly referred to as Outback Areas.

The area of our ethnographic study encompassed the USG and nearby contiguous lands of the Outback Areas where renewable generators are located. The whole area is attractive for renewables generation because of the abundance of both elevated and flat land, a sunny climate, and favourable coastal wind patterns blowing off the Gulf. A large proportion of the land surrounding the three USG cities is taken up by extensive pastoral properties, some of which now host large-scale solar and wind generators.

The USG cities have long been associated with heavy industry, with the region historically referred to as 'the iron triangle'. Today, iron ore mining and minerals processing (steel, lead, and other metals), construction, port operations, and the railways remain important employment sectors across the three cities, which also supply workforce and services to other mining industries operating further north in the Outback Areas. Port Augusta was the centre of the state's coal-fired electricity generation until 2016, when its last power plant was shut down.

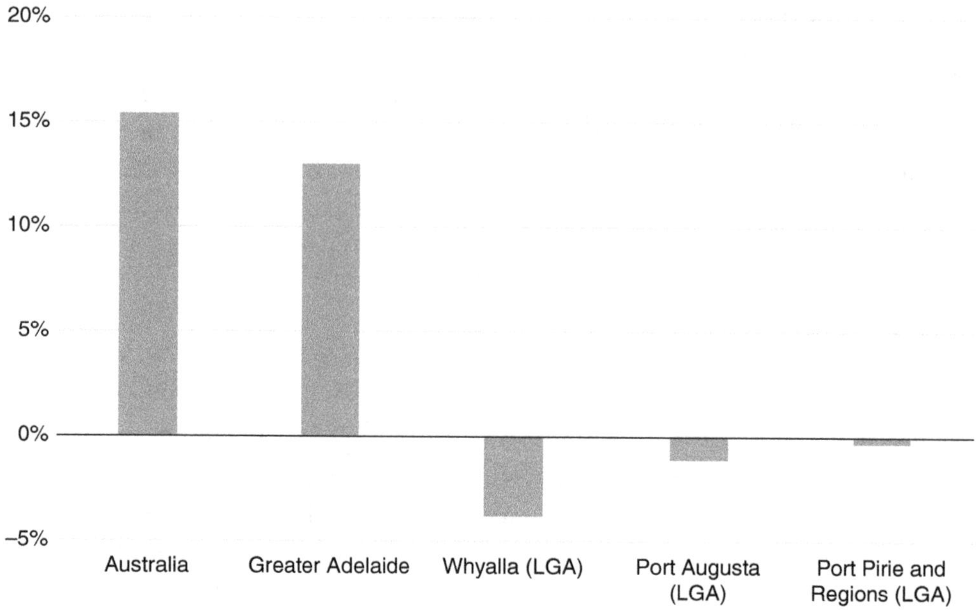

Figure 5.1 Comparative population variation: 2011–2021 (Source: 2021 and 2011 ABS Census QuickStats)

As Figure 5.1 shows, the population in all three cities is gradually shrinking in contrast to Greater Adelaide and Australia, leading to multiple social and economic concerns among residents and local government and feeding the hope that new industries like renewable energy may provide population stability, revitalisation, or even growth.

The USG population consists of a relatively high number of Aboriginal people, especially in Port Augusta (20.4% of the population). Whyalla (5.5%) and Port Pirie (4.7%) are also high compared to Greater Adelaide (1.7%) and Australia (3.2%) (ABS Census Quickstats 2021). Since colonisation, Port Augusta has become a service centre for northern remote populations, but, for many thousands of years, the 'top of the gulf' has been an important First Nations meeting place. It is a place that features in many interconnected Indigenous songlines splaying out across the continent, and traditionally served as a place of trade, inter-marriage, diplomacy, cultural ceremony, and more.

The two groups of Aboriginal Traditional Owners, Barngarla and Nukunu, whose lands encompass most of the USG (shown on Map 5.1) now have full native title recognition (see later section). However, most Aboriginal residents in the USG are not from these local Traditional Owner groups. Many were forced there from other Aboriginal nations due to colonial dispossession and others have moved there, both as long-term or temporary residents, for work, to visit family, and to access services such as health and education. Due to continuous

Table 5.1 *USG: the 'regional gap' by comparative demographics*

	Australia	Greater Adelaide	Whyalla (LGA)	Port Augusta (LGA)	Port Pirie and Regions (LGA)
Population	25,422,788	1,387,290	21,244	13,829	17,282
Income					
Median weekly household income	$1,746	$1,548	$1,188	$1,275	$1,063
Variation: 2011 to 2021	+42%	+40%	+27%	+29%	+38%
Work					
In the labour force	61%	62%	55%	53%	52%
Highest level of educational attainment					
Either year 10, 11, or 12	30%	32%	40%	39%	40%
Bachelor degree or above	26%	26%	10%	10%	9%
Health					
Reported long-term health conditions	36%	36%	44%	46%	46%

Source: 2021 and 2011 ABS Census QuickStats.

disadvantage, these populations statistically tend to experience poorer health, higher unemployment, and lower educational achievement than non-Aboriginal residents.

In general, the USG is a region of significant socio-economic disadvantage compared to the majority population of South Australia. Table 5.1 compares some key indicators of what we have termed the 'Regional Gap'. While this is a limited statistical profile, it shows some marked disparities with state and national indicators.

Figure 5.2 shows that unemployment in the USG is strikingly higher and more volatile than the rest of the South Australian economy. The pattern of price cycle downturns, restructuring, routine maintenance schedules, expansions, and closures of large industries contribute to the higher unemployment periods, as do the peaks and troughs of short-term construction work associated with renewable installations and other large infrastructure projects.

The geography, history, and socio-demographic profile of the USG present a picture of socio-economic potential for renewable energy investment, but also some limitations related to population size, income, employment, and education levels. These form both part of the legitimacy context for the new industry's operations and also present as challenges, as our case studies will analyse.

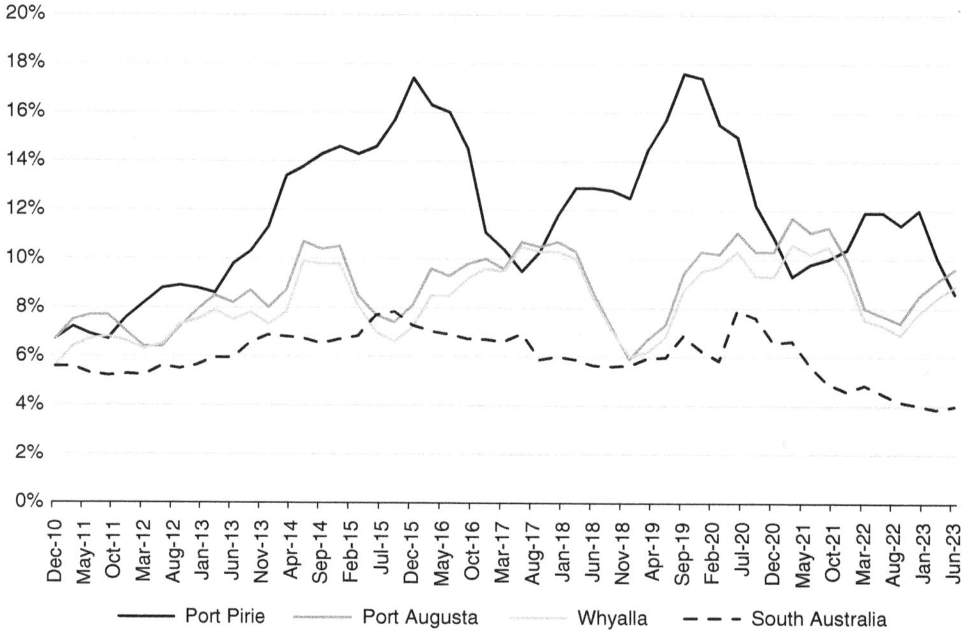

Figure 5.2 Unemployment rates in the USG: December 2010–June 2023. South Australian data for the June 2023 quarter is calculated from April/May data only. Source: Jobs and Skills Australia, Small Area Labour Markets (June 2023); Labour Market Insights, All Regions (ABS SA4) – Population by labour force status, April 2023.

Repower Port Augusta Campaign: Early Hopes for Renewables

In 2012, the Port Augusta Power Station was the largest private employer in the town, but signs of its demise were beginning to appear. In February, the Playford B generator was put into 'reserve storage' by owner Alinta Energy. This left just Northern, the largest and youngest generator, as the state's only remaining fully operational coal-fired power station. Then the Federal Labor Government announced a 'contract for closure' plan, which explored the notion of paying the nation's dirtiest coal-fired power stations to close (Combet 2012), with Port Augusta near the top of that list. Even after this idea was politically abandoned (Cullen 2012), it was followed by the introduction of a carbon price.

It was around this time a small group of Port Augusta residents mobilised to start a grassroots community campaign they called Repower Port Augusta (RPA). They wanted to prepare for an eventual closure of the local coal-fired power stations and take up the opportunities presented by new large-scale renewable technology that could offer a 'just transition' and did not pollute their air.

In 2012, Beyond Zero Emissions (BZE), an independent climate change think tank, released *Repowering Port Augusta*, proposing a plan to transition coal power workers to concentrated solar thermal (CST) power generation. The vision '... would enable South Australia to become a world leader in renewable energy, and Port Augusta would become an iconic global hub for baseload solar power generation' (Beyond Zero Emissions 2012: 4). BZE proposed a phased installation of six CST generators, plus extra wind power, replacing the existing coal generation (around 760 MW). CST generators employ a central tower surrounded by an array of giant mirrors (heliostats) directing the sun's heat up to the receiving tower tip, capturing the sun's heat for electricity production.

Concentrated solar thermal technology outlined in the BZE report won the support of Port Augusta residents. CST had been identified to be an excellent match for the existing workforce profile, plus offered potential for employment growth in the manufacturing of heliostats. CST also made sense to local people who were already 'energy literate' due to their long power-generating experience, in that CST provided storage and dispatchable power. People were also drawn to it because it presented an opportunity to redefine their 'dirty, working-class town' image.

Repower Port Augusta participants recalled to us how, from March 2012, a loosely organised committee of about 12–15 volunteers met monthly. A community organiser was also funded through not-for-profit organisations: Australian Youth Climate Coalition (AYCC) and later by the solar energy activist group Solar Citizens. They began with a Facebook page and a RPA Alliance with local government, businesses, health organisations, unions, and climate and environment groups, expanding the resourcing and impact of the campaign. Soon after, in July 2012, they organised a 'community vote' that asked residents whether they wanted to see the coal power station replaced with gas or solar thermal. Around 30% of the town participated (3,500 respondents), 98% favouring solar thermal (ABC News 2012a).

Gary, a mechanical technical officer who had worked at the power plants for more than 20 years, was positive about the compatibility of skills between power station and thermal solar workers: 'solar thermal ... had that long duration storage and a good number of jobs in building and maintenance [and] a range of more interesting jobs compared to PV and wind' (Interview Gary September 2022). Other RPA committee members reflected that the conviction and enthusiasm that welled up in those early months was premised on seeing 'opportunity ... employment benefits without the negative health and environmental price tag' (Interview Gary November 2020).

The RPA campaign momentum continued to grow. In early August 2012, South Australia's Energy Minister met with solar thermal lobbyists and announced his 'fact-finding tour' to Nevada, to visit the world's largest operational CST plant, owned by US-based company SolarReserve (ABC News 2012b). In mid-September, the AYCC organised about 30 of their members from across Australia to 'Walk for Solar' in a 300 kilometres two-week walk from Port Augusta to Adelaide, culminating in a rally at Parliament House attended by more than a thousand supporters, with speeches by the Port Augusta Mayor and the State Premier, Jay Weatherill (Channel 10 2012).

Repower Port Augusta activism, combined with proactive state renewable energy policy, was part of a 'renewables rush' in the USG. Renewable energy prospectors and international companies, including a few CST companies, began vying for grid-accessible land. Solar Reserve was one, and as an industry leader at the time, they secured the RPA campaign's support with their 150 MW CST Aurora project. This proposal involved a 110 metre tower for a concentrated solar power plant to be located 30 kilometres north of Port Augusta, on pastoral lease land.

In 2015, the power plant and coal mine owner, Alinta, announced they would close all operations within three years, as they had become 'increasingly uneconomic' (ABC News 2015). Shock and despair rippled through the community. An Australian Manufacturing Workers Union official declared that the closures will 'absolutely devastate Port Augusta' (ABC News 2015). The announcement surprised Gary and his fellow employees at the plant. Just a few months before, Alinta publicly stated that the Northern power plant would operate until 2030, and management told workers they were reviewing options, including reduced production. As it played out, the feeder coal mine was closed in nine months and the operating power plant was closed within a year, with the loss of around 400 regional jobs. Without the 520 MW Northern power station capacity in the grid, it was not long before a severe electricity crisis ensued. Triggered by a cyclone in late 2016, an expensive state-wide power outage resulted, leaving some regions out of power for nearly a week. This was then followed by a hot 2016/2017 summer of electricity supply shortages, worsened by some of the highest electricity prices in the country and fierce mismanagement accusations from the national government.

Over the preceding five years, the RPA campaign had expanded beyond the Port Augusta community to a state and then national initiative. The electricity crises had the political leaders frantically looking for options, and RPA was ready. Amidst this political crunch, Solar Reserve's CST Aurora project secured a vital 20-year Power Purchase Agreement (PPA) with the SA government in August 2017 stipulating local (SA) supply quotas, including heliostat production. The project development application was lodged in September 2017 and swiftly approved by January 2018.

However, in April 2019, the company suddenly announced cancellation of the project (ABC News 2019), being unable to secure sufficient project finance in time to meet the state government contractual requirements. The poor performance of the Nevada generator, which Gary was tracking, likely underpinned the situation. Stunned landowners and other residents heard the news via a media announcement from the SA Minister for Energy on 5 April 2019 (ABC News 2019). Solar Reserve dismissed all local employees, ceased community liaison, and filed for bankruptcy later that year following technical issues at its Nevada solar plant (REVE 2020).

Solar Reserve's abrupt exit demoralised the community. Louise, mentioned at the beginning of the chapter, worked for the company and shared RPA's aspirations for a renewables-led revitalisation of the town. She recalled:

> Solar Reserve was such a wonderful opportunity for Port Augusta.... I was very disappointed in how it was handled, from the community perspective, because the company's approach was just 'shut the place down, that's the end of it'. I feel like we workers were left, as community members as well as employees, to bear the brunt, not getting any information apart from the instruction that we weren't allowed to say anything. (Interview Louise November 2020)

The social legitimacy of energy transitioning, inspired by the grassroots RPA campaign, evaporated with Aurora's collapse. Shock and disappointment were enduring, because the campaign had been associated with broader ideals for renewable energy transition based on community leadership, grassroots democracy, climate action, just transition, and regional justice. The sense of social power to create desirable new futures crumbled as locals came to terms with the lost prospect of Aurora's generator.

Although these hopes proved to be ephemeral, several other projects did proceed in the region. As we examine further in this chapter, large-scale renewable energy initiatives across the USG have had diverse trajectories, some failing to progress like the Solar Reserve project. Others have reached operational status, with varying experiences of engagement, contestation, and support in the various social groups of their host communities.

Repower Port Augusta supporters reflected to us on the value in the longer-term positive implications of their campaign effort, as expressed by Gary:

> I think the value that we've ultimately brought by what we did was that we did make our community more aware of the whole issue and that's a good thing. I feel that we helped encourage and to some extent educate a couple of State energy ministers on renewable energy ... we probably should be encouraged that at least we've got almost as much peak capacity coming from Port Augusta now or will have when the Port Augusta Renewable Energy Park is online, as we had when both Playford and Northern power stations were running. (Interview Gary September 2022)

The RPA initiative was an unusually forceful community campaign that emerged in a critical period of ruptured expectations and uncertainty about the future of a socio-economically disadvantaged regional town undergoing abrupt closures and downsizing of major industries. The collective vision of a new industry centred on solar thermal generation could not be realised, with blockages in corporate financing and technological innovation that went far beyond the scope of regional activism and the available levels of state incentives. As this example shows, outcomes in renewable energy transition are uncertain, though it does provide lessons that can be drawn upon to expose the potential and the limitations of community action, which can be a resource to influence subsequent developments.

The Arrival of Renewables, State Regulation, and Local Government

The process of appropriation that is required for renewable energy capital to thrive requires the application of state power through regimes of planning that broker and sanction changes to land use and socio-economic relations in host communities. The aim is for developments to proceed with a sufficient measure of social legitimacy reinforced by regulatory compulsion. In Australia, the locus of power to activate renewable development rests largely in the planning and regulatory regimes of the state governments. South Australia, as a leader in Australia's energy transition, illustrates some of the tensions in play.

In South Australia, large-scale renewable energy projects are subject to regulations and assessment procedures defined by the state's *Planning, Development and Infrastructure Act (2016)*, which was progressively rolled out between 2019 and 2021, during much of the period of this study. The State Planning Commission is the delegated independent body for guiding state and local governments, as well as developers and host communities through the application, assessment, and decision-making process. It operates the State Commission Assessment Panel (SCAP), responsible for assessing all state developments, including large-scale renewable energy applications. The SCAP makes referrals (recommendations) to the Minister for Planning and Local Government in accordance with the *Act (2016)*.

In an interview with personnel from the Department of Energy and Mining, we were informed that most large renewable energy developers access the development process via a 'pre-lodgement service' that enables them to ascertain whether there are any obvious blocks in the approval process (Interview June 2021). Once given the preliminary all-clear, they need a government agency willing to sponsor their project for the purpose of providing 'essential state infrastructure' and to recommend their development application via the Crown Development route (PlanSA 2023). While this form of application is subject to the same public scrutiny as other development applications, it is preferred because the final decision made by the Minister cannot be appealed. This minimises the risk to developers of costly delays but is a dead-end road for third-party objectors.

Large-scale renewable project developers focus on extracting quality resources as efficiently as possible for maximum profit, while fulfilling government regulatory requirements, maintaining social legitimacy in host communities, and offering the best returns for shareholders. In our interviews with representatives of companies operating in the USG, they explained three main criteria they consider when assessing a prospective project location for a development application:

First, there is a focus on the natural resource potential. This concerns the availability of large land areas and the quality of the wind or solar resource, initially 'discovered' and measured by prospectors and technical contractors who negotiate exploratory site access with landholders, Aboriginal title holders and government land authorities.

Second, developers seek to assess constraints, in terms of issues that are likely to arise and delay or prevent the project. Delays cost money and are 'a hassle'. These can include: the number of landowners in negotiations; various objections or concerns raised by landholders and neighbours; local community attitudes; Native Title and other land use restrictions; cultural heritage sites; or matters of environmental protection, such as the presence of populations of threatened species. Any of these can delay a project for months (or even years) either before or during the application process, due to the requirements of additional data collection and liaison with land managers, local government, conservation authorities, Traditional Owners, Aboriginal Corporations, neighbours, and other residents.

Third, developers seek to establish the availability of transmission line access and connection. National authorities must be assured via complex data modelling that any new project will not jeopardise grid stability and that sufficient grid capacity exists before permits will be granted. 'Connection is king', as more than one developer remarked. The further a project is from the grid, the bigger the project scale required to absorb the transmission construction costs. The open access system in South Australia means projects tend to cluster competitively around existing high-voltage networks. Connection approval is a lengthy, multimillion-dollar process that may impede financial close on a project, and can sometimes extend into the construction phase, without any certainty of an optimal outcome. One developer explained:

> It's a very detailed technical modelling process that looks at any kind of scenario that you could imagine in the grid: low demand; high demand; different generators dropping on and off; how's your plant going to respond to ensure that it's not going to impact the grid at all. It's the same for any new project trying to connect. And it's constantly evolving because we're responding to new information and rule changes playing into it. (Interview Developer November 2020)

As the Port Augusta Aurora project illustrated, financing too is uncertain, often until late in the development process. Offtake or PPAs with energy-intensive organisations or companies, ideally for the life of the generator, are usually a crucial part of securing financier support on a project. But most governments no longer award PPAs. Contracts for difference and other price guarantees/options contracts, and underwriting corporate PPAs, are much more common. Often companies need 50–100% of the project capacity secured via an offtake agreement to be able to proceed to construction. The competition among companies to secure these agreements was strong during the period of this study, intensified by reverse auctions, and as one developer remarked, 'they are hard to come by' (Interview Developer December 2020).

The uncertain pattern of project development in the USG is evident in Appendix A, in the delays from development application lodgement to operational status for many projects, with some cancelled and others failing to progress past development approval to construction, over extended periods. Several projects continue to claim they are 'active' but what that really means is they are still running the developers' gauntlet and holding out hope they will get over the line, as this project manager remarked: 'We just have to make a calculated assessment of the risk. Honestly, that's all we can do. If everyone waited until they had full certainty, we'd never transition the grid to renewables ever, so we have to just take a bit of a punt' (Interview Developer November 2020).

The lack of visible progress at sites, after the media fanfare of a project approval, contributes to the waning of residents' interest. Inactivity is confusing and frustrating and can lead to mistrust for groups hoping for benefits, such as small business contracts or community grants, or it can exacerbate opponents' anger and suspicion. One developer told us that they try to manage expectations by limiting communication to news of important project milestones, while others 'make a lot of press releases, talk their project up, and then, after a few weeks, silence!' (Interview November 2020). USG residents recall multiple renewables proposals that were announced with fanfare and then faded away: the Cultana salt water pumped hydro, the Goat Hill pumped hydro, the Solarstor graphite CST project, and the Playford battery were often mentioned.

Local government representatives shared with us their feelings of frustration about inert projects that could be contributing to economic well-being. One manager said of the delays: 'That is one of the realities that the community doesn't understand. You know, these applications come in and it's quite a long process before approval is given. The community expects to see something on the ground soon, whatever it is' (Interview November 2020).

An USG Mayor explained his perspective:

> One of the things that's frustrating me the most is the longevity between the application and the build. Because we run over thirteen per cent unemployment here, and we could really do with the jobs. There are different phases of work throughout the build period,

so it'd be about two years of work. Then it would drop down to nothing, but two years of having something, rather than nothing, is a life-changer for some families. And it might break the cycle, for example, of them being on the dole [welfare]. (Interview USG Mayor November 2020)

During the course of the field research, we observed a significant reduction in new large-scale solar, wind, and battery development applications in the USG after late 2019, and that several projects, despite planning approval, were not proceeding (see Appendix A). There were a number of reasons for this.

In the USG, the major northern high-voltage line was historically established as part of a state-owned network distributing electricity from the Port Augusta coal-fired power stations to the Adelaide metropole and manufacturing hubs in the south of the state. In 2020, when the government announced a new 860 kilometres 330 kV high-voltage interconnector route connecting South Australia to New South Wales known as EnergyConnect, large-scale renewables investment attention in South Australia moved eastwards to the Mid North region, the new transmission capacity, and the interstate energy export potential.

Also, from 2019, all new energy developments over 5 MW connecting to the state's electricity network were required to seek approval from the South Australian Office of the Technical Regulator (OTR). As a part of the approval, every renewables project in South Australia is required to provide either a synchronous condenser or a battery to meet the OTR grid stability and frequency control requirements (SA State Planning Commission 2019: 7). This increased costs for projects and together with possible curtailment requirements or grid connection problems meant the return on investment became even more uncertain. In restricting the accumulation potential of new projects, this regulation was at odds with the state's pro-active investment strategy, but on the other hand protected government politicians against the electoral liability of severe power outages for which voters have a long memory.

Local Government: Energy Investment Trumps Local Tax Revenue

In interviews with local government representatives and managers, they expressed hope and some optimism that new energy industries would benefit the region. But there were common threads of concern about depleted population (see Figure 5.1) and revenue, ever-increasing 'red tape' and the shrinking authority of councils. Development planning decisions have shifted from individual councils to regional planning boards and, for large projects, to the SCAP and the Minister for Planning. The passing of the *Hydrogen and Renewable Energy Bill* in December 2023 continues this process (Government of South Australia 2023). In our interviews with local government personnel, the most common and contentious matter was the failure of the state government to allow land taxes to be levied on renewable energy installations.

Under South Australia's *Local Government Act* (1999), local councils can determine the taxation (within limits) levied on land within their jurisdiction, except for land used for electricity generation and transmission. This exemption originated from a time when electricity was almost entirely public owned by the state government but remained in place after privatisation in the late 1990s. In lieu, the government nominated selected allotments used for power generation and required them to pay an indexed sum to the local council, commonly referred to as debenture agreements. By 2016, the agreement relating to the Port Augusta power stations was worth nearly $500,000 annually to the Port Augusta Council.

As renewable energy developments sprang up on different land allotments, regional councils found they could not levy rates above the existing zone type, despite the change in use. Several regional councils dealing with these challenges joined forces to fund a report that could provide insight as to how councils might best proceed and what changes might be requested of the state government. The report found that local communities were 'subsidising electricity generated for use across a broader (and national) marketplace' and that anticipated industry growth 'will place an increasing burden' on local councils and communities (Legatus Group 2018: iv).

Regional councils have lobbied for state legislation changes that would allow Councils to recover payments directly from renewable energy generators according to a formula, subject to indexation, similar to the historical debenture agreement system. They argued that other states like Victoria and Queensland required companies to pay local taxes and that regional communities in South Australia were being short-changed and unfairly burdened. The Energy Minister we interviewed in 2021 was cautious about changes to the relevant Act, signalling possible complications and unintended consequences, including the prospect of creating a development deterrent in a competitive Australian and international market and the possibility of higher power prices for all. The status quo remains despite a change in government in 2022. Thus, the state has supported a new form of appropriation while relieving companies of land tax costs and depriving poorer host communities of public revenue to local government.

Case Study: Crystal Brook Energy Park

It is against a background of regional socio-economic disadvantage, project cancellations and delays, as well as local government revenue restrictions, that the social legitimacy of projects is called into question and that new processes of appropriation and accumulation come into play. Crystal Brook Energy Park is a contentious case that brought all parties into contradictory relationships of conflict and alliance in a small rural locality, with long-lasting effects and an unsatisfactory outcome for the developer.

Crystal Brook is a picturesque town of 1,322 people located 29 kilometres southeast and inland of the city of Port Pirie, and within the Aboriginal native title determination area of the Nukunu people (Australian Bureau of Statistics 2021). The town's surrounding hills are at the southern tip of the Flinders Ranges, and it is a hub for grain-growing and grazing farms. An energy park development proposal there in 2018 illustrates the role of the state planning authority in processes of ecological appropriation through an unusually intense conflict within the host community.

The wind resource potential of the low hills, 3.5 kilometres northeast of Crystal Brook, had been prospected in the early 2000s, but development discontinued in 2012 (Neoen Australia 2018a). Seizing an opportunity in 2016 after the power station closure, the previously involved landholders initiated their own competitive tender process for a wind and solar farm on their collective properties and were approached by 20 companies. French energy giant Neoen won the tender with a proposal to develop and operate a wind, solar, and battery energy park.

The project was approved to go forward as a Crown Development, and Neoen submitted its application in May 2018 with a proposal for 56 wind turbines (up to 125 MW capacity), a 150 MW solar farm, and a 130 MW/400 MWh battery, with substation and associated infrastructure. A 50 MW hydrogen production facility was also flagged but not included in the application.

The application proceeded to the mandated community consultation stage in June 2018. The public submissions revealed intense opposition among landowners of the Beetaloo Valley, a residential area of 41 dwellings (population 75, 2021 Census) in smaller acreage properties to the north of Crystal Brook. Land use in the Beetaloo Valley is mixed: small farms and orchards, grazing, and a growing number of vacation rentals, promoting the valley as 'the gateway to the Southern Flinders Ranges', a hub for scenic walking trails, and a place of 'picturesque views with peace and quiet'. The residents' Beetaloo Valley Association energetically organised information and meetings and provided support for many objections to the development. Some Crystal Brook residents joined the objectors, and the campaign was fortified by the participation of anti-wind farm organisations, notably the Waubra Foundation, one of whose leading members had a residence in Beetaloo Valley. Opponents also mounted a strong campaign at PPRC, which despite having limited actual influence on the planning decision voted in principle against the project in June 2018. USG councils, unable to access any tax revenue from renewable installations, have little incentive to resist strongly objecting residents. The council's stance was perceived as validating opposition sentiment towards the project (Clarke 2018).

Wide-ranging objections outnumbered supporters in written submissions, with almost all focusing on the wind farm component. The strength of opposition led to the planning assessment panel authorising a second round of verbal submissions at public hearings in October 2018. The day before the hearings, Neoen made a significant change to its application, announcing a halving of the number of turbines (from 52 to 26) but with an enlargement of their size and height (from 185 m to 240 m). We learned in interviews that this decision, which deprived some landowners in the partnership of a significant amount of long-term lease income, added to the tensions over the project.

Neoen's concession did not appease the opponents, who reiterated their objections to the wind farm, encompassing the full range of energy conflicts in Becker and Naumann's (2018) typology in overlapping ways. The solar farm attracted very few objections. Some written submissions raised concerns about the technology: the Waubra Foundation and similar groups opposed wind turbines because of generalised concerns about negative health impacts, specifically low-frequency noise and vibration. Locational conflicts were the most prominent, with objectors identifying construction and operational risks as well as landscape views, including the cumulative visual impact of the large number of wind farms already in the Mid North region.

There was frustration at the scale of the wind farm, with one objector stating 'you wonder where it's going to stop if people don't stand up and say no'. Others were concerned at the impact on the ambience of the nearby heritage park and forest conservation area. Property values were a concern, with one resident stating 'our views are one of our greatest selling points'. Other locational issues included the impact on a major long-distance hiking trail that passes through the area; setback distances from turbines; impact of land clearing and turbines on flora, fauna, and biodiversity corridors; aviation interference including aerial agriculture and firefighting; and telecommunications interference. Broader identity conflicts were strongly combined with locational concerns: objectors singled out threats to the public perception of the bucolic landscape, including the attractiveness of Crystal Brook and environs as places to live and the importance of restful natural surrounds for tourism marketing, especially in the Beetaloo Valley.

The above concerns played into distributional conflicts around the benefits of the project. Some were critical of host landowners whose 'blind greed for wealth' threatened asset values and business operations. Recriminations were vented over the company not caring about these threats. Concerns were expressed about the pressure on local services and infrastructure during the construction phase.

At the same time, there were intense procedural conflicts through the consultation stage and beyond. Many objectors complained that the company had never contacted them or that meetings and other communications had been inadequate.

There was a pervasive sense of a predetermined outcome and lack of transparency about decision-making in many of the submissions and in some of our interviews. Neoen (2018b) responded to the verbal submissions with counter-evidence of varying detail and quality, and there were some undertakings for further investigation on some issues (e.g. telecommunication interference from the turbines).

Some Nukunu native title holders had talks early on, but they were not able to pose a public challenge as the forms of freehold and other tenure of the lands in and around the project site had automatically extinguished any native title rights. Despite the concerns raised in submissions, there was considerable support for the project in the community. Host landowners predictably came out in favour, and Neoen's undertaking to voluntarily fund a community grants programme was welcomed, with one resident commenting: 'We have 26 community organisations here. It's very difficult for them to raise funds in this small town' (Interview November 2020). An anonymous survey by the local newspaper, while not methodologically robust, indicated 83% of respondents in favour of the development (Clarke 2018).

Residents we met remarked that, in a small community like Crystal Brook, many cross-cutting social ties inhibited people from publicly voicing strong opinions on a divisive issue. Avoidance is the most common tactic for dealing with pervasive local conflicts. Repercussions can be indirect. For example, one host landowner told us that when the owner of his agricultural supplies firm spoke out against the project, he changed suppliers.

The planning assessment panel deliberated on the application, including all submissions and responses. In its April 2019 report, the SCAP generally weighed in favour of the Neoen, while alluding to the need to address issues of land use, zoning, and layout (Webb 2019). The report noted that the project was compliant with the relevant legislation and guidelines, endorsing it for 'further review and decision' by the Minister for Planning. The project was approved in August 2019.

Residents we interviewed the following year, who opposed the project, were still distressed over the outcome. Several people commented to us that they were closed out of opportunities to participate after the submissions to the public hearings. They regarded the endorsement of the project as unjust, personally traumatising, and damaging to the civic culture of the community. Residents who supported the project, council officers, and company representatives we interviewed also reported how hard the process was for them personally. At the time of writing, the project is still in limbo and Neoen is building a far larger energy park in Goyder, with prime access to the previously described EnergyConnect interconnector with NSW. Industry observers have suggested to us that without the hydrogen production facility and a reduced number of turbines, the Crystal Brook project is not financially competitive.

The Crystal Brook Energy Park is an unusual case for the USG. Strong opposition from a minority of residents in the host community sharply impacted the project's trajectory and community relations. The planning assessment panel had a more prominent role in this locality than in any other project in our study. Because of the public challenge to the 'socio-ecological fix' sought by investment capital, the conditions of production of Neoen's venture were brought into question. The energy park site is part of a wider ecology – sky, air, waters, landforms, flora, fauna, cultural heritage, and dwellings. They are perceived largely as commons, already reserved for others' diverse livelihoods and ways of life. But the state deemed this ecology to be of greater value for an energy park than for the other values that residents placed on it. In the interests of decarbonisation, state-wide voter popularity, and private accumulation, these other local values were rendered peripheral by the state's regulatory regime of ecological appropriation: a satisfying outcome for many but calamitous for others.

Legitimacy and Contestation: Land

The socio-ecological relations of renewable energy industries depend on place-fixed assets for generators and transmission. Large-scale wind and solar farms add another layer of commodification and biophysical transformation of land and ecologies, continuing the capitalist appropriation of ecologies since the beginning of the colonial state. At the same time, the burden of fossil fuel extraction is lessened as the inexhaustible sources of sun and wind are harnessed for the production of sustainable energy.

The state of South Australia is almost 1 million square kilometres. Large tracts of land in suitable sites are attractive to private renewables investors incentivised by the state. This land is, however, already owned and used under state laws, or claimed under different laws by First Nations people. Aboriginal Traditional Owner groups hold title to 22% of the land area, under the state's Land Rights and Land Trust Acts, and the area of native title lands is increasing as Aboriginal claimant groups achieve successful determinations.

Pastoralism (grazing of sheep and cattle) is a dominant industry in the northern, more arid zones of South Australia. In the mid- and southern regions, grain growing prevails. Over 40% of land in the state is held as pastoral leases from the government, with a variety of tenure arrangements (South Australia Department for Environment and Water n.d.). Pastoral leases (termed 'stations') average tens of thousands of hectares, with some properties covering hundreds of thousands of hectares (South Australia Pastoral Board 2016: 5–6). The large landholdings and expansive open spaces of the USG and Outback Areas have high-quality resources for both wind and solar generation. Depending on terrain and size, wind farms require land areas of a few thousand hectares (such as the Port Augusta Renewable Energy Park (PAREP), with 50 turbines installed on 5,000 hectares of flat land adjacent to the Spencer Gulf) to much more (such as the Lincoln Gap wind farm, with 101 turbines on 20,000 hectares of hilly terrain in the Outback Area southwest of Port Augusta). Solar farms intensively use smaller areas of land and occupy flatter topography compared to wind farms. For example, the Bungala Solar Farm near Port Augusta has a generation capacity of 275 MW from over 800,000 panels on 800 hectares of gently sloping land.

Large-scale renewable generators are land-intensive assets that impose significant changes on the landscape during construction, and wind farms especially continue to be a prominent landscape feature during operation. We encountered some instances where aspects of these installations were publicly contested during the development assessment process, most notably the Crystal Brook Energy Park. There are also examples with very few or no formal objections, and others that received a positive reception. As we explain in the latter sections, not all opposition played out in the planning system or other public fora, instead existing as a quiet but ongoing resentment or tension.

Host Landowner Arrangements

Benefits to landholders who host large-scale renewable generators can be considerable, providing stable revenue and business diversification, with lease contracts covering the operational life of the generator, of 30-plus years. At the time of our study, the older and smaller wind turbines (1–2 MW) each typically provided income of $5,000–$10,000 per year to a landowner. Larger wind turbines that generate 5–7 MW each were worth about $25,000 each per year for host landowners, but fewer units are installed. Solar farm leases are usually based on an annual per hectare amount for the life of the generator. Leases to solar operators were returning annual payments between $300 and $3,000 per hectare, depending on resource quality and expected generation profits. Some land lease arrangements included a percentage of the wholesale price of power sold to market, and annual income for other related infrastructure and access arrangements.

Wind farms in the USG do not generally require trade-offs with other forms of farm productivity, as the post-construction turbine footprint is relatively small, and the land can continue to support livestock grazing. Solar panels occupy most of the leased land allocation, and, in the USG, secondary use for farm production (agrivoltaics) is not in practice. As farm sizes increase under pressure to achieve economies of scale, the scope for large-scale renewable energy as a form of stable economic diversification increases. The owner of an USG agricultural supplies company in a cropping area explained the pressures for farmers:

> Land prices have gone up threefold in the last few years because farmers must keep expanding. The equipment is getting larger and more expensive. They need to spread the risk and need larger areas to get the return and make it economical. These days, a top harvester would cost $1 million, but it's only used for six weeks. Plus you need a new truck, new storage, a bigger seeder and a bigger tractor. You can't buy small machinery anymore. But the cap on returns is commodity prices and that affects solar farm lease negotiations. (Interview Agricultural Supplier November 2020)

In an interview with us, the Australian Energy Infrastructure Commissioner observed that the renewables developments with the least public complaints were those with 'large acreages and a small number of landowners, and no neighbours, no towns or other settlements' (Interview Commissioner January 2021). He remarked that there were very few complaints from Western Australia or South Australia where most projects met these criteria.

Lincoln Gap Wind Farm

The Lincoln Gap Wind Farm is one such development in our study area, located on steep hills that stand out from surrounding flatter country, 27 kilometres south-west of Port Augusta. Host landowners Bruce and Julie have a pastoral holding of several hundred thousand acres, not far from the high-voltage transmission grid. Bruce recounted how they became leaseholders for the wind farm in 2004, a year of severe drought for farmers:

> Mark was the spotter, and he'd been over to Alveston, and his first thoughts were to go to the coast where the wind was, but he was just driving back through here and he thought, 'Hills, powerlines. Huh. Wonder who owns that?' He tracked us down. He was spotting for a couple of Irish lads actually, the Darcy brothers, who came and lived in Australia, and they saw the potential. Very quickly got us into an agreement, put up some wind masts and got a development approval [in 2006]. But it took a long time to get through the next stages. They wanted to sell the project. I took a lot of prospective developers up there to have a look over the years, probably ten dozen. (Interview Bruce November 2020)

The Darcy brothers finally made a competitive sale, and then the contract was on sold several times without any progress towards construction. In 2016, Nexif Energy, a partnership of developer Nexif and investment firm Denham Capital, acquired the lease contract. Due to the long period since the original approval, technology changes, and financing delays, in 2017 they successfully submitted an application for extension of the approval with increased turbine size and generation capacity. The original 59 turbines were replaced by 101 taller ones and a battery, in a three-stage 468 MW project. Bruce and Julie said that with no other landholders, neighbours, or settlements nearby, the lease negotiations were relatively simple. However, even with such an easy process, problems emerged down the track.

Julie recounts that they had to take a strong stance over the construction company's plan for the main access road to run right by their house, a stance that delayed the development by 6 months:

> It was nearly due time that they were going to start working and they said, 'We're going to have to use this road'. And I said, 'No, you're not, you're just not going to use it'. Yeah. There was already quarry sand out of the creek and a good track around the corner there. I said, 'Well why don't you use that?' 'OK, we'll look into that'… And there's always little things. There's always something. (Interview Julie November 2020)

When we interviewed Bruce and Julie in November 2020, they had just received their first payment from Stage 1 of the wind farm, which had begun full generation earlier in the year. Bruce remarked: 'So it's sixteen years since we signed the agreement that we got paid, but the rate was set back there. We've got a sixteen-year-old product. We have to wait for "practical completion". I'd say it's like that for most landowners'.

The risks of long delays and financial uncertainties became apparent when Senvion, the turbine supplier used in the project, went into receivership in 2019. After some delays, Nexif turned to another manufacturer, Vestas, to complete Stage 1. Landowners do not bear any direct costs of the project, and Bruce and Julie have remained optimistic and are now enjoying the income from both Stages 1 and 2 of the project, with Stage 3 well underway. Their land, much of which is under freehold title, is also sought after by solar developers, and a pumped hydro project was also on the cards for several years. They are open to more renewables expansion but have become quite selective about the proponents they are willing to work with and are careful to assess the feasibility and risks of proposals. Bruce surmised: 'That's why we're not too excited about being a developer, because you've got to "stick to your knitting". Now, I understand sheep and it returns dollars. [Renewables] is a complete new business. You've got to understand your limitations, don't you?'

In terms of the typology of energy conflicts outlined by Becker and Naumann (2018), there have been no significant local conflicts associated with the Lincoln Gap wind farm and the project has enhanced the value of the landowners' property. The Aboriginal Traditional Owners, the Barngarla people, may have had a view, but the freehold land title limited the challenges. Other USG projects presented different profiles of land use arrangements and local alignments of support and opposition.

Port Augusta Renewable Energy Park (PAREP)

The PAREP was initially developed by Irish-owned DP Energy, a 'front end' developer that works in several countries, identifying sites, securing land lease options, and obtaining planning and connection approval before on-selling (DP Energy n.d.). Phase 1 of the project was bought in 2020 by Spanish transnational energy company Iberdrola. Construction began on the 217 MW wind farm and 110 MW integrated solar farm in October 2020, both of which became operational in 2022.

The site is adjacent to the eastern side of the Spencer Gulf. Beginning 8 kilometres south of Port Augusta, it extends southwards on both sides of the national highway towards Adelaide. Its coastal location catches early morning and evening winds from the Gulf, complementing hours of solar production and peak grid demand. It is semi-arid pastoral land, zoned Primary Industrial, in smaller holdings of several thousand acres used for running small numbers of sheep and cattle. Four landowners are leasing to Iberdrola with options dating from when DP Energy prospected there in 2009–2010. The three smaller landowners have six to eight turbines each, while the largest landowner has the remainder of the project's total of 50 turbines. They have a 'twenty-five plus twenty-five' year leases, with a fixed fee per turbine (reputedly $25,000), a percentage of generation, and continued grazing rights.

In interviews, we heard grievances from Nukunu Traditional Owners, who were granted native title rights subsequent to DP's land negotiations. They complained that the developer did not carry out representative consultation processes as required for Indigenous Land Use Agreements to award compensation, resulting in particular families or factions receiving benefits that should have been distributed to all. Additionally, some of the final turbine sitings are culturally sensitive. For these reasons, the PAREP location has been the subject of internal tensions within the Nukunu community, not aired in the public domain or through the planning system.

Host landowners, as is usually the case, had few complaints about construction impacts. One whom we interviewed mentioned 'they had cut up a lot of land', more than he expected, but new fences and rehabilitation were part of the contract. Dust was clearly a problem, with the company delivering hay when natural cattle feed became too dusty to eat. However, 'Neighbours haven't said anything. A few people don't like it, but there are no major complaints' (Interview May 2021). Nearby there is a small settlement of residents on allotments of a few acres each, on land zoned Rural Living. We encountered no strong complaints from these project neighbours, although some mentioned the rumble of large truck movements, construction dust, and ecosystem damage as concerns.

The most vociferous objections to PAREP came from some of the residents on the western side of the Gulf who faced the prospect of their views of the iconic Flinders Ranges (see Map 5.1) being marred by the turbines. A feature of the western coast of the USG are the shacks along the shoreline, originally makeshift structures built by railway workers for weekends and holidays. Two hundred and eighty shacks line the beach, some now replaced by large suburban houses or modified by extensive renovations.

The Shack Owners Association and individuals, including some tourism operators, made written and personal submissions to a Planning Commission hearing on the development application in Port Augusta, but the approval went through in 2016. A development variation application in 2019 to increase the tip height of the turbines from 150 m to 185 m triggered a new round of objections. The shack owners' submissions claimed that the digital images submitted by DP Energy to the Planning Commission minimised the size of the turbines and that the visual amenity of the views they enjoyed would be ruined. Residents in an adjacent suburb raised anew their concerns about noise and dust. All those in opposition stated that they did not oppose renewable energy or wind turbines *per se*, but they did not want them to go there.

Most local government representatives and many business leaders supported the project as essential to the future of the town. Strong objections came from some councillors: 'Everyone that is local knows that the Flinders Ranges changes colours from late afternoon to sunset, a magnificent view. After 60 years of looking at the hills, to have this put in place, the planning people need to be removed' (Green 2019: 1). The Mayor took a conciliatory stance: 'With all these projects there are positives and negatives. It is a fantastic thing that could occur to promote employment in our city, but in the back of our minds we are taking into consideration the questions that the community is asking about it as well' (Green 2019: 1).

The conflict had the character of the locational and identity conflicts as outlined by Becker and Naumann (2018). The visual amenity of a beautiful landscape, noise and dust exposure to nearby residents, and the identity of the town as a tourism destination framed by the silhouette of the Flinders Rangers were all at stake, not to mention the threat to land values that some shack owners raised. Project delays required technology upgrades that revived some residents' opposition. Eventually, the bigger turbines were approved. Unlike Crystal Brook, Port Augusta locals reported to us that the community divisions, between the 'shackies' and project supporters seem to have faded in the bigger township of Port Augusta.

Native Title and Renewables in the Upper Spencer Gulf

As briefly mentioned in the earlier case studies, there are two main Aboriginal nations in the USG. They are Barngarla, whose determined land stretches north from the top of the Gulf and across the Eyre Peninsula to the west, and Nukunu, to the east of the Gulf (see Map 5.1). Like other First Nations in Australia, they have suffered dispossession, displacement, massacres, and other violence from the time of first contact with Europeans, intensifying when settlers were awarded unceded land grants in the mid-nineteenth century by the South Australian colonial government. People maintained connections to their country, working on pastoral properties and residing on allocated reserves and on the fringes of towns. Both the Barngarla and Nukunu people have been prominent in the struggle to regain rights to their lands.

After many years of uncertainty pressing their claims under the Native Title Act, both Barngarla and Nukunu have had a series of successful determinations since 1998 (Federal Court of Australia 1998, 2016, 2021). Their native title rights now extend across the USG and into the Eyre Peninsula and Outback Areas, some in high-value resource areas for wind and solar generation, as well as associated battery and hydrogen production sites and electricity transmission infrastructure. Across two decisions in 2016 and 2021, the Barngarla people were awarded native title claims stretching to the north, west, and south of Port Augusta, including two-thirds of the Eyre Peninsula and the industrial cities of Whyalla, Port Lincoln, and Port Augusta (Federal Court of Australia 2016, 2021). Similarly, the Nukunu people were granted native title across two decisions in 2019 and 2022, encompassing 50% of the Port Augusta LGA and land to the northeast, east, and south of Port Augusta, including along the eastern border of the Spencer Gulf and inland to the east over the southern Flinders range (Nukunu Native Title Claim 2019). These determinations took decades to accomplish and involved much contention about overlapping claims between neighbouring native title claimant groups (Roberts and Aeria 2022).

However, only a small proportion of the land within the mapped Barngarla and Nukunu determination areas is native title land. Native title may be exclusive or may co-exist with other forms of land title or tenure. Native title is extinguished where there are already exclusive rights to the land, such as freehold title (Commonwealth of Australia 1993). Within these restrictions, Native Title corporations have rights regarding the use of the land and access to it, including any proposed future changes to the land by other parties. One lawyer in the field gave us a simplified summary:

> Native title is complicated. But the simplest way of describing it is that it's an interest in land which comprises somewhere between 50 to 100 per cent of the land that it attaches to, depending on the other tenure on the land. If there's other [pre-existing] tenure on the land like a pastoral lease or some kind of non-exclusive lease or licence, that tenure will continue to operate, so it won't affect that. But any additional tenure granted or any change to the existing tenure is what's called a future act.
>
> So fundamentally once someone tries to enact new tenure, such as a licence or an easement or a lease, or converting something to freehold or the equivalent, then that is where native title rights apply, and the [native title holders] have a right. … The Native Title Act has created these property rights and now you get some engagement with companies who have to pay their fair share [as it's a change to existing tenure]. (Interview Lawyer August 2022)

The Barngarla and Nukunu determinations mark some important changes during the period of our research. They are an important new base of Aboriginal property rights in the expanding field of energy developments in the region (and in many other parts of Australia). Aboriginal relationships to land embrace deep and ancient attachments to unique places, are an intrinsic part of social and spiritual identity, and are far removed from the commodity relations to land in the commercial sphere (Moreton-Robinson 2015). Conflicts about Aboriginal land are always conflicts about identity, connection and dispossession, and the sacredness of a location. Specific renewables projects where Aboriginal lands have played an instrumental part in project development are discussed further below.

Project Development on Aboriginal Lands

Successful registration with the National Native Title Tribunal is an arduous process, requiring claimant groups to demonstrate 'continuous and unbroken connection' with the claimed land (or waters). It also requires the formation of a single entity, a Registered Native Title Body Corporate (RNTBC), also known as Prescribed Body Corporate (PBC), for the management of native title matters for the group (AIATSIS n.d.).

For most of the currently operational large USG renewable energy projects, the Barngarla and Nukunu people were not yet native title holders. While developers are required to consult with native title claimants as well as title holders, the former are at a disadvantage compared to recognised native title groups, who are more empowered in negotiations that take place through registered Native Title corporations.

As earlier case studies touched on, developers were ignorant or neglectful of Traditional Owners' rights and unappreciative of cultural complexity. Our interviewees reported that consultation was often superficial, non-existent, or with self-declared individual representatives, fuelling internal disputes and distress. Companies anxious to progress their applications were intentionally or unintentionally advantaged by this fragmented and sometimes conflictual state of community relations, which our interviewees felt was further exacerbated by a lack of proactive government support for native title applicants and holders in the development process.

We heard forcefully expressed complaints about the poor outcomes from Aboriginal people dealing with renewable energy projects, especially where native title determinations had not been finalised and strong corporate boards were not established. The legal wrangles, among families and with developers, can have long-lasting effects on those who did and did not consent to projects. Traditional Owners expressed intense suffering from the capitalisation of nature and the lack of respect given to their unceded sovereignty of ancestral lands, the harms to ritual places, spiritual connection, and songlines, as well as the unfair and poorly regulated procedures that do not ensure 'free, prior and informed consent' from the whole group of Traditional Owners with native title rights.

One Aboriginal Corporation official lamented the 'corporate lowballing' that characterised negotiations for native title compensation for renewable energy projects (Interview Official August 2022). The potential for significant economic benefits is compromised, and the process is experienced as a kind of recolonisation. Another Traditional Owner said to us: 'I've had no good experience with renewable energy yet. Just a lot of indignity and pain. At least with coal, they didn't pretend to be something they were not. People got paid' (Interview May 2021).

The experiences of Aboriginal people with developers, whether in the broad category of Traditional Owners, or as native title claimants/holders, raised the most complex pattern of conflicts in our study. There are entangled tensions and disputes of a distributional, procedural, locational, and identity nature, as we outline in Chapter 1, drawing on Becker and Neuman (2018). However, these problems rest on a more fundamental historical and endemic conflict stemming from the colonial state's doctrine of terra nullius, its appropriation of Aboriginal Lands, and its failure to recognise sovereignty or make reparations.

The legal recognition of native title in 1993 has been the most significant step towards justice. Since the determination of native title on Barngarla and Nukunu lands in the USG and nearby Outback Area, these nations have registered corporate bodies mandated by the Native Title Act. They are now in a better position for more representative negotiations of rights and benefits accruing from the use of their land for projects, but the legacy of former dispossession and displacement persists, as we discuss in the next two examples.

Bungala Solar Farm

Apart from Lincoln Gap and PAREP, the only other large-scale renewable energy installation that was under construction and became operational during the period of our field research in the USG was the Bungala Solar Farm. Bungala is a 275 MW photovoltaic installation in two stages, 10 kilometres to the northeast of Port Augusta, spanning the boundaries of Port Augusta Council and Outback Authority jurisdictions. The semi-arid land is located close to the high-voltage transmission line and consists of low-grade pastoral land owned by Bungala Aboriginal Corporation (BUAC), within the Nukunu native title Determination Area. The BUAC, based in Port Augusta, has no traditional cultural connection with the land or the solar farm, nor native title rights. BUAC is set up 'to provide employment opportunities for Indigenous people in Port Augusta and the surrounding regions'

(Bungala Aboriginal Corporation n.d.). Aboriginal Corporations are legal entities set up under federal law, with certain restrictions and exemptions, for various purposes of Aboriginal and Torres Strait Islander welfare and advancement as well as governance of native title determinations (Office of the Registrar of Indigenous Corporations n.d.).

A Crown Development application was lodged by the original developer, Melbourne-based Reach Solar Energy, in 2015 and was approved in 2016. The project was then purchased by Italian energy company Enel Green Power in 2017. Construction by lead EPC contractor Elecnor began in April 2017 and was completed in November 2018. There were numerous problems, including dust storms from the denuded site afflicting the health of a nearby Aboriginal community and town residents, and the death of a sub-contractor employee, crushed by a pile-driving machine. After construction ended, a run of serious technical difficulties and national energy regulator curtailment restrictions prevented the project from becoming fully operational until September 2020.

Prior to the solar farm proposal, the Bungala land had been subject to overlapping native title claims by Barngarla and Nukunu groups. We heard about Barngarla and Nukunu dissatisfaction when the state government transferred the land with freehold title to BUAC in 2005. Both groups repudiated the right of BUAC to own the land, maintaining that the government had bestowed freehold title in error, as the land was previously under the control of the Aboriginal and Torres Strait Islander Commission (a body abolished by the Federal government in 2004). While BUAC retained freehold title, the land was determined to be within Nukunu's successful native title determination in 2016. Nukunu claimant organisations signed an agreement with Enel for heritage site protection and some financial compensation.

Despite legal and financial advice during negotiations, the CEO of BUAC told us that in hindsight they did not get the best lease deal from the original developer, Reach Solar. BUAC signed a 30-year lease with CPI increase and a major review after 15 years. A requirement to employ 30% Aboriginal workers, managed by an employment agency in Whyalla, was written into the contract. All subcontractor procurement had to agree to employ Aboriginal people. Retrospectively, the CEO believes that BUAC should have gotten a much better deal, including long-term security of contracts, a percentage of power revenue, and/or cheap power. With transmission lines extending through the land, BUAC also receives nominal additional income from the South Australian electricity transmission corporation, Electranet.

Bungala Aboriginal Corporation found the lead (EPC) contractor Elecnor to be inexperienced with recruitment and employment of Aboriginal workers. Subcontractors were not supportive and Aboriginal employment quotas were not reached during most of the build. The CEO recounted that when Enel Green Power bought Bungala, there was a marked improvement in support for Aboriginal employment. In the last few months of construction, 40% of the workforce was Aboriginal. An Aboriginal-staffed mobile food cart to feed workers on site was also developed as a way to facilitate more female employment and benefit.

The Bungala Solar Farm shows how property and land rights are a foundational element in achieving inclusive and equitable outcomes for Indigenous people in the landscape of place-fixed assets in the renewable generation industry. However, outcomes are variable as these rights exist in complex social fields where good outcomes from effective corporate policies, legal support, and Indigenous organisational strength can be undercut by competing claims, discrimination, economic exploitation, profit maximisation, weak regulation, and government disregard.

Yoorndoo Ilga Solar

With native title determinations largely settled across the USG, native title holders can fare better as new projects develop. Barngarla Determination Aboriginal Corporation (BDAC) has been particularly advantaged by this recognition as Whyalla and lands in the Eyre Peninsula are now the focus of energy industry future acts including hydrogen production and export projects, feedstock solar generation, and transmission line upgrades. The 2016 Barngarla native title determination included land north of Whyalla where AMP and EPS Energy have development approval (January 2024) for a 330 MW solar farm and 250 MW battery. BDAC made an Indigenous Land Use Agreement with the South Australian government in January 2020 to surrender and extinguish their native title on this land in exchange for freehold title, thus enabling full commercial negotiation rights with developers (Cultana Industrial Estate Crown Land Transfer 2020). BDAC was then able to negotiate a lease agreement with the developers on favourable terms. Planning approval was granted in January 2024.

Yet to start construction, the project has already brought Aboriginal owners and the developers into a relationship of mutual benefit celebrated by both sides, through a tightly negotiated legal agreement that the Barngarla board chair told us he hopes will be a benchmark for the future of renewable energy on native title lands. One of the ongoing benefits of renewable energy projects for BDAC has been their insistence that project developers fund its legal spend equivalent to theirs. Other Aboriginal groups have approached these complex legal negotiations with only a few thousand dollars in hand for a minimal level of advice, resulting in poor outcomes. For their first solar project, the BDAC board was therefore able to engage a large legal and technical team to 'break ground' on a fair agreement. BDAC, the developers, and the legal team received a Commendation in the Community category at the South Australian Premier's Energy Awards in 2022. In a media release, BDAC stated:

> This project, which covers almost 700 ha of land and is worth around $300 Million, will provide significant jobs and reduce energy costs to regional SA. It was only possible, because BDAC through our lawyers were able to negotiate with the State Government to allow for 700 ha of native title land to be handed back to us the Barngarla People in freehold, to allow the BDAC to enter into a lease to allow the project to progress. This was also only possible, because BDAC & Normans [legal firm] were simultaneously able to negotiate with the proponent companies EPS Energy and Amp Energy to cover the costs of the transfer and for Barngarla to take an equity stake in the project as part of the lease agreement.
>
> The Lease means that not only are BDAC the landholders and landlords for the entire project, but we will also have a shareholding in the project, making us – the local First Nations People for the area – part-owners of one of the largest renewable energy plants in the State. This is also the first time in SA that land has been handed back to First Nations People for a commercial purpose, making it a pioneering project in terms of future development. (Barngarla Aboriginal Determination Corporation 2022)

In a video made for BDAC members, the chair of the board, Jason Bilney, explained the significance of the 2021 land use agreement with the developers of the Yoorndoo Ilga Solar Farm, which he described as a 'landmark deal'. He emphasised that the requirements of their native title claim had brought the three formerly conflictual families of Barngarla into a cooperative framework for common ends through their corporation. His words were full of optimism for the future: 'Our grandparents and great-grandparents have fought for our native title, and now what we are doing is setting our children and grandchildren up. BDAC has turned into a very sophisticated Aboriginal corporation, and we want to continue in this way with new generations coming in, and new ideas. Watch this space' (BDAC and MAV Media SA 2021).

Jason sees the conditions of the lease as game-changing, not only for Barngarla but also for other native title groups negotiating with energy companies:

> The biggest thing of all is to push that envelope [in legal agreements]. We are landlords for this project. We take share ownership. Others will take a full lump sum, we want it over the life of the project. So we get an income for every year from when the project starts construction until it actually shuts. It could be fifty years, it could go longer. So income every year from those projects… We have Key Performance Indicators in it as well… There are jobs provided, and school-based apprenticeships… Once we get established with the income, we'll set up a trust… We've sort of set a benchmark [with the first solar project]. It's important having a board switched on, let alone having the community involved. In the meantime, there are a lot more projects, solar, wind farms and hydrogen projects which are happening in the USG. Now we've set the boundary of one, we can set the boundary of a lot more within the area. (Interview Jason August 2022)

Barngarla are among the native title corporations that have assertively pursued their rights in negotiations with government and developers. A 2024 front-page media article stated the various negotiations for renewables, hydrogen, green iron, and other developments will create revenue streams making the 'Barngarla Determination Aboriginal Corporation the first in Australia to be worth at least $1bn, particularly without any government funding' (Starick 2024). All the while other groups have achieved only poor outcomes in what can be time-consuming and expensive legal fights, especially where native title is not yet determined. One lawyer working in the field explained:

> We see that a lot of groups just go, 'Oh, it's too difficult, we'll just let it go through'. But then they don't get the benefits that come from [a good agreement], which is really problematic. If the companies pay, it will make them less reckless. It's effectively bringing a negative externality into the market, and puts the cost burden on a wealthier body of the community, who pay more than they have in the past. (Interview Lawyer August 2022)

This is a very different perspective on developer contributions to the Energy Minister quoted earlier. For him, creating any extra costs to developers (such as levies to local councils) was considered a disincentive to investment and raised the spectre of higher costs being passed on to voters' electricity bills.

Nukunu native title holders still express much angst over the unsatisfactory lease arrangements on their lands, and where access is denied to land on which they now have native title rights. One elder reflected on the improvement since the determination:

> It's all in place now. We've got the Prescribed Body Corporate [Nukunu Wapma Thura Corporation], and legal support, and consultants who can do this and do that for us. And we are very thorough in what we go through. We don't want stuff taken away that we've fought so hard for over all these years now; that's our attitude, and we stick strongly together on that. (Interview Elder September 2022)

The most contentious negotiation since the determination of Nukunu native title has been the proposed Baroota Pumped Hydro project, on a reservoir site in the vicinity of an area that's very sacred to Nukunu people and includes an endangered species. We were told that the Nukunu native title corporation ceased negotiations with the developer despite increased offers of financial compensation, as the board decided the project was an extreme danger to one of their most important dreaming places. In their view, no monetary compensation or cultural heritage agreements could offset the risks.

The only surviving original Nukunu claimant, Lindsay Thomas, expressed the optimism and sense of empowerment that the native title has brought: 'We can build and go forward – we know now where we stand on country and we can get to work on our plans to make a healthy country' (quoted in Roberts and Gillian (2022)).

Social Legitimacy of Energy Transition in the Upper Spencer Gulf? Current Realities and Future Expectations

Heavy industries have dominated growth in the USG economy since the mid-nineteenth century. USG residents have long navigated an unwritten social contract where governments or companies establish industrial operations to exploit resources there, attracting workers and a multi-generational resident base for enduring industrial enterprises. To a point, these communities have accepted the degradation of the environment, some hard working conditions, and risks to their health from industrial pollutants, as with coal ash from coal generators in Port Augusta, the lamina of red iron dust over the town of Whyalla, and lead in the air and soil of Port Pirie.

While not all residents are employed in these industries, the whole of the community advantages that USG communities are familiar with them delivering have historically included many long-term good jobs, traineeships and apprenticeships, affordable housing, grants and sponsorships for community organisations and events, local tax income to councils, and stable relationships with decision-makers. That is not to say there are no conflicts: workforce reductions for example are a recurring event and are vigorously contested by unions and advocates for the region. As the power station closure underlined, job security has declined along with opportunities for many residents including youth and Aboriginal people.

These benefits of industrial life have contributed to the attractiveness of the region for generations of working people who have a sense of their valued roles in the wider economy. Some of the benefits (like housing, community infrastructure, and stable cohorts of local managers) are part of a bygone era, while others continue. Nonetheless, these types of community conditions, while smaller now, have provided an historical framework for socio-economic vibrancy and set a precedent for the industrial social contract.

In Pursuit of Socio-economic Vibrancy

The power station closures, the failure of the RPA campaign, and the periodic crises of the lead smelter, steelworks, and mines have significantly ruptured the expectations of the social contract. Mark, a local who runs one of the largest crane-hire companies in South Australia, reflected on the demolition of the power station stack in 2018:

> [T]hat was a sad day for the town. You think when I rocked up here [in 2001] with two old cranes there was a railways with 350 people building locomotives, and refurbishing locomotives, and building rolling stock. At the power station there were 400. The coal mine with another 150, 200. And a pretty vibrant movie-making district as well. (Interview Mark November 2020)

There are mixed views about the prospects for a secure USG future, and uncertainty about the promise of renewable energy. All local council and regional development personnel spoke of plans to foster employment, diverse economic growth, and social vitality through tourism and new industries. One council officer explained: 'Our council wants appropriate development and job opportunities. We see things like renewable energy projects as assisting us in changing negative perceptions of the industry base of Pirie – to potentially one of being a green, or a greener-tinged, industrial city' (Interview Council Officer November 2020).

In-migration is an important part of councils' growth strategies as the Port Pirie Mayor explained: 'We have been promoting the fact that, because of the high-speed connectivity and all that, you can live here and still do your work and have this relaxed lifestyle' (Interview November 2020).

While the restrictions of the Covid pandemic during 2020–2021 fuelled a population shift to regional areas Australia-wide, the USG cities continued the trend of overall population decline except for one quarter of growth in Port Augusta (Australian Bureau of Statistics 2022; Stephens, Cansdale, and Forbes 2022). Renewable energy projects, which bring employment during short periods of construction, have not yet generated any ongoing population growth.

Former power station employee Gary explained that Port Augusta in particular had been the target of a number of failed 'town-saving' schemes promising to help revitalise socio-economic vibrancy. These included a petrochemical complex proposal and a 1999 Teletrack online racing project backed by the South Australian government, focused on international gambling audiences. He went on to analyse the impact of the crash of the solar thermal project: 'Then Solar Reserve falling through means people will be really quite sceptical, and I personally think, reluctant to put any skin in the game as far as even speaking in favour of new projects let alone actually doing anything to make them happen' (Interview Gary September 2022).

The forms of disengagement are subtle. Gary explains:

> It's going to make it difficult to get overt public support because sadly, it's very easy to put chin on something, 'That'll never happen', and if it falls over, 'Yeah, I told you so'. And if it does come and you don't have to do anything, you just shut up and take your job there if you want and all is good. (Interview Gary 2022)

Despite the quiet scepticism that Gary and others have identified, energetic local government groups cast a positive light on the economic future, echoing state government declarations about the growth potential of renewable energy:

> The USG region has historically been a significant economic contributor to the state and nation through its natural resources and industrial strengths. Over the last decade, the regional economy has diversified to capitalise on those natural resources to become a

> significant renewable energy 'powerhouse'. Planned public and private sector investment in hydrogen power, supported by significant infrastructure projects over the next five years, will further establish the USG region as one of the most important nationally in meeting emissions targets. (Spencer Gulf Cities 2022: 1)

The three USG LGAs incorporated in 1998, forming the USG Common Purpose Group (known as Spencer Gulf Cities) to 'share information, jointly implement initiatives, provide a united voice and work with Government, industry and other stakeholders in the interest of improving the long term sustainability of the region' (Spencer Gulf Cities n.d.: 1). They are planning for 9% population growth in the decade, an ambitious target when set against the trends of population decline (see Table 5.1). The group seeks to leverage opportunities from new regional development including the anticipated construction pipeline of renewables jobs in Whyalla and Port Pirie, the hydrogen generation and export industry, battery recycling, the Southern Flinders National Park, and mining the old Port Augusta Power Station ash dam for cement manufacture.

In state planning and policy statements, specific acknowledgement of the region's socio-demographic gap is absent, as is specific information about how renewable projects will address disadvantage in terms of basic life indicators like health, education, employment, and income. Rather, the political stakes are waged around the promise of cheap, abundant, and reliable clean energy for the whole state and demonstrating growth in Gross Domestic Product (GDP) and productivity measures.

In our interview with Minister Dan van Holst Pellekaan, we raised the issue of social and economic values for host communities when the government makes a long-term PPA with electricity retailers, as initially happened with Solar Reserve and subsequently happened with the $1 billion dollar contract with Zen Energy in October 2020 (Government of South Australia 2020). As part of the 10-year agreement, Zen Energy and Simec Energy were required to maintain their head office in Adelaide, instead of moving it to the eastern states and were contracted to build a large solar farm adjacent to the steelworks in Whyalla and a utility-scale battery in Port Augusta. To date, the head office remains in Adelaide, but neither of the projects has progressed to construction, and Simec and Zen have dissolved their partnership.

> INTERVIEWER: So there are no kind of requirements on Zen to bringing knowledge, education, research … or other kinds of obligations they had to fulfil as part of that contract?
>
> MINISTER: [T]o be honest, we could have put anything we wanted in there but then the cost of electricity just goes up and up and up so you've got to find that balancing point between spending more taxpayers' money on electricity and making sure that what you get is worth it versus if you try and put too much in your bucket … you just have to spend way, way more for the electricity. (Interview Minister April 2021)

The Minister went on to explain the priorities that were driving state policy on renewables development:

> If you just want to get your electricity price down, you can do that much more easily and not worry about emissions, not worry about other things. If you just want to get your emissions down, you can do that much more easily but not worry about reliability or the cost. [In addition, we want to] decarbonise highly energy intensive industries. We are getting cleaner energy and cheaper energy into all of these processing opportunities.
>
> Working on all three of those things and requiring that all three of those things move in the right direction simultaneously is absolutely fundamental to my mindset and where I am going. So then linking that back to what you're asking about, if throughout South Australia we are delivering cheaper, cleaner and more reliable electricity, every business, every household, every brand-new processor, everyone is advantaged. (Interview Minister April 2021)

This thinking is in tension with the strivings of the USG community, local government, and other regional advocates who actively pursue local outcomes. As one USG resident commented to us, 'there is too much *big* thinking'. The additional state-wide electricity bill savings of '$10 on average for households' later announced by the Minister (Filatoff 2020: 1) seem modest compared to other forms of social and economic values that already socially disadvantaged host communities might receive from agreements, such as with Zen Energy.

Elusive Jobs

In a region with high unemployment and closure or downsizing of major industries, attracting new 'clean' industries that deliver good jobs is a priority in the USG. New jobs figure prominently in the announcements of politicians when projects are approved, as seen in media coverage of projects associated with the Zen Energy agreement:

> 'We're committed to building what matters, using our budget to drive jobs and growth to recover from COVID-19', said the State's Minister for Energy and Mining, Dan van Holst Pellekaan of the dual projects to be delivered by Simec Energy Australia, which will create some 810 new jobs during construction and require around $40 million worth of local goods and services. (Filatoff 2020: 1)

The policy mechanisms to translate the government's prosperity vision into local conditions are unclear. We asked the Minister about the low numbers of local workers on USG projects that we had observed, the lack of long-term work many complained about, the difficulty of accessing necessary and timely skills training, and the issues with work readiness and generational unemployment. He agreed that it could be improved but argued that trainers and job seekers should take the long view rather than be negative, that the projects 'will keep coming' and construction work will continue to be available (Interview Minister April 2021).

Speaking about Port Augusta, Mark, the local crane-hire business owner, said: 'This is just a welfare town really'. He went on to say: 'I sort of get a bit embarrassed sometimes. You know we have delegations of different mining companies come through sponsored by local business organisations, or SACOME [South Australia Chamber of Mines and Energy], and they start to talk about capability, and you can just see everyone's face drop, because the capabilities just aren't here' (Interview Mark November 2020).

Government analysis of business scale and skills for renewable energy supply chains confirms Mark's observation. One Regional Development Australia (RDA) officer, Claire, explained that there were no regional businesses that met the criteria for a 'Tier 1' construction business. The USG had one private business classified as Tier 2 (Mark's crane hire business with 170 employees), but most businesses were Tier 3, such as gravel suppliers and emergency electricians who could service renewables projects. Claire observed that skills training is also patchy.

Ruling party politicians usually tend to pass over the capability gap, the practical barriers to accessing the opportunities, and the manifest socio-economic disadvantage of the USG when spruiking the economic virtues of new development to a region. The limited potential shown by renewable energy projects to provide the sort of stable jobs and workforce benefits of the old industries is glossed over: one industry is just like any other. As the local state government Member of Parliament (and Energy and Mining Minister), Dan Van Holst Pellekaan said when we interviewed him: 'So some industries are still here, some are not. Some new ones that weren't here are coming. So in my mind at least it's all industry, it's just some of them are transitioning' (Interview Minister April 2021).

Each of the large-scale wind and solar projects under construction during the period of fieldwork required a large onsite workforce, with numbers and duration of jobs varying over multi-year construction periods. As well as providing some employment to locals (mostly short-term contracts or casual work), the construction projects have attracted hundreds of workers from interstate as well as other parts of South Australia and overseas, many brought in as part of contractors' crews. One manager at a labour hire agency in Whyalla commented:

> The six hundred jobs announced for the proposed Cultana Solar Farm in Whyalla are only for four months work. One hundred and sixty of those jobs will come from elsewhere and require specialised skills. You can't expect to get solar farm engineers from the local community. We will try to get about three hundred plus local workers. I can guarantee though that most local hires won't stay on site for longer than 31 days. (Interview Labour Hire Agency April 2021)

He explained to us that unemployment benefits are cut off after 31 days of work. As a result, he said that people are unwilling to risk losing their safety net and do not want to go through the time-consuming process to re-register for benefits. He continued: 'Also, many people don't want to work 11-hour days in dust, rain, hail or shine. We will definitely have to recruit [overseas] backpacker labour. There is a cost to keep hiring new people, so employing locals who won't stay is more expensive than backpackers'.

In an effort to keep local workers, the agency was 'trying something different', offering to support those who stayed for the entire work period to enrol in a technical college electrical certificate to get recognition and upgrade work skills.

Recruitment agency managers must adapt to the unpredictable and short-term nature of renewable energy project employment demands. Dave, the Port Augusta branch manager of a major national company, explained that a lot of their work is labour hire. Some workers choose to work in short-term jobs, especially farmers in off-season work periods, semi-retired workers, and those on unemployment benefits as outlined earlier. Dave explained that renewables projects are 'a revolving door of staff'. Apart from weekly or day workers, contracts are generally for 3, 6, or 12 months. This pattern of employment is common to all large renewable construction jobs, which he contrasted to mining, railways, and the former power plant jobs, all longer term with a career trajectory.

> The wind and solar farm projects are working on high pressure timelines, six days a week. It's not like mining where processes are planned and structured and regulated. In mining operations, there are three to four people with the skills to do every job. In renewables construction they don't have the back-up up staff who can fill in when someone is sick. It makes the whole implementation process vulnerable. We come in when there are urgent requests to provide the backfill. (Interview Dave May 2021)

In these circumstances and where there is not a full labour procurement plan when companies arrive, the major contractors become reliant on agencies and local employees to assist with urgent recruitment needs, whether it be through social networks, liaising with local hospitality staff in pubs who 'know everyone', or depending on the 'free' services of recruitment agencies (all methods we observed). Dave described an example where he recruited 20 concrete truck drivers for one renewables project in the region, and then created longer-term contracts by coordinating them rotating between three project sites. In such situations, Dave provides the service of aggregating work across multiple sites to create a good employment contract for workers. The goal of better workforce outcomes through cross-project coordination has been an aspiration of training providers and a recommendation of the state government's Industry Advocate, but has been difficult to achieve except via the *ad hoc* initiatives of local recruitment agency personnel with detailed project knowledge (such as Dave).

Another employment services manager, Lisa, highlighted the general difficulties of finding jobs for unskilled entry-level workers on renewable energy projects. The long project lead times, the vague timing of labour requirements, and short periods of construction made it hard to activate a large pool of workers at the right time, and coordination to maximise workers' employment across projects was a challenge. There was no time to train people on the job and renewable energy companies generally do not want to invest in traineeships or apprenticeships for short-term jobs. She remarked that it is often disheartening for local jobseekers when major renewable energy projects recruit outside the region. Attempting to bridge the capability gap, Lisa had established good relations with some employers and there were collective plans for a new industry-accredited course in solar farm construction (Interview Lisa February 2021).

Upper Spencer Gulf personnel in training agencies that we interviewed regarded the opportunities in the renewable energy industry as limited. Mandy was the Business Development Manager for a SA-owned private registered training organisation with offices in four sites including Port Augusta. She explained:

> We specialise in high-risk work licences to operate heavy equipment required by SafeWork South Australia and we have 40 units of training on offer. The vast majority of our business is in the mining industry. We've tried to engage with the renewable construction projects but contractors bring their own crews who are already competent. It's understandable, its efficient and they can keep their staff in contract work. Like TESMEC [an international trenching and cable laying company with an Australian office] – it brings all workers from interstate or overseas, except the site administrator. (Interview Mandy March 2021)

Mandy surmised that 'renewable energy doesn't have the best reputation around town as an industry offering good reliable jobs'. She had seen no spike in training for renewable energy projects, and no reason to set up training with no demand. She said, 'If locals do get a job and want longer term work, they have to join a crew and travel. But, it's a job, and the pay is good, with allowance for working away conditions'.

Most tradespeople and other local businesses that we spoke to preferred to keep their local clientele rather than seek short-term contracts on the renewables projects. Obtaining the extra skills to apply for tenders was seen as a lot of effort for an uncertain outcome or long-term value. Maintenance work on operational generators was preferred, but generally kept to a small part of business revenue. For example, Mark, the crane-hire business owner, stopped tendering for construction work on renewables projects because of the uncertain timelines, short-term contracts, and fast-changing technology, preferring to undertake wind farm maintenance work, limiting the renewables share of the business to about 15%. Supply chain businesses like gravel and water supply, road works, and equipment hire were the most likely to incorporate renewable project work but avoided high dependence on it because of the short-term nature of contracts.

The projects have stimulated some new enterprises and expansion. In 2019, two local nurses set up a now flourishing clinic focusing on construction workers' health and safety screening, testing, and other services (Stellar Health Solutions n.d.). National construction materials supplier Boral set up agencies for gravel and concrete in Port Augusta for the duration of the intense period of renewables construction that has now come to an end. Boral has more permanent agencies at Whyalla to service iron ore mining and steelworks. And in 2020, a local business-woman established an USG-specific online jobs hub 'connecting local employers with jobseekers, making it easier to live and work in the vibrant USG of South Australia' with renewables companies as regular clients (Upper Spencer Gulf Jobs n.d.: 1).

Hospitality and food retail business are the largest beneficiaries of the waves of non-resident workers on the projects, with real estate businesses and landlords also benefiting. However, the influx of construction workers puts pressure on housing in host communities. Real estate agents spoke of hiked rental prices and charities told us of an increase in rental stress and homelessness during construction peaks.

Aboriginal native title corporations have been able to leverage jobs and other benefits for members as part of land use agreements with renewable companies building on native title land. Their negotiations include meaningful long-term jobs and training for the duration of generator operation as well as construction. The Chair of Barngarla Aboriginal Determination Corporation, Jason Bilney, said:

> It took us twenty-one years to win our native title. Our grandparents and great grandparents have fought for our native title and now to see us doing what we're doing is setting our children and grandchildren up.... It's all coming together as one ... bringing the three communities together.... Having more young people involved. The hardest part is yet to come.... We want to get all these young people into these jobs ... get out there and work on the country, be cultural awareness officers, be part of the monitoring team. Be out there and work with the companies. (BDAC and MAV Media SA 2021)

For politicians and agencies seeking to improve employment outcomes across the region, even a few ongoing jobs offered by operational generators are seen as a plus. The Energy Minister explained to us:

> If any project came here and gave us twenty extra sustained jobs, we would say thumbs up, that's great. It's about how can we get more really. To my mind, this is a huge opportunity for the USG. It's actually about putting in more industries other than just the generation, so making this an energy hub beyond just what the wind and the sun and the available land and things offer. So there is software and hardware and there is manufacturing of all sorts of things. There are a lot of other parts that come into energy. (Interview Minister April 2021)

Residents do not dispute the value of these small numbers of long-term jobs, even if few in number. But most of the accessible local jobs in renewables are lower skill and/or short-term during construction and there are only a handful of ongoing operational roles, while the renewable companies are seen to accumulate value for shareholders over decades.

The deferral of more sustained and widespread local economic benefits to other future industries that can leverage the 'cheap green renewable energy', as expressed earlier by the Minister, presents yet another iteration of the government's prosperity promise, like a 'benefits mirage', serving to further undermine the social legitimacy of the transition. It is symptomatic of the political discounting of the pressing difficulties of everyday life that ripple through social relations in the USG host communities, despite the heady promises of energy transition. As one Council CEO commented to us, 'We have received very little value from these projects so far'.

Conclusion: A Questionable Transformation

For more than a decade, South Australia has been at the forefront of Australia's clean energy transformation and today the state is a world-leading jurisdiction in electricity system decarbonisation, as discussed in Chapter 2. In 2023, over 70% of South Australia's electricity consumption was produced by renewables, largely wind, followed by rooftop solar, with increasingly frequent days of 100% renewables supply (OpenNEM 2024). As of recently, substantial state and private investment has been flagged for hydrogen production, power plants, and storage in Whyalla, as well as a hydrogen export hub at nearby Port Bonython (Shteyman 2023).

The state's admirable statistics are important for global climate change outcomes as well as achieving government emission reduction targets. But energy transformation is socio-ecological as well as technological. This understanding has directed the inquiry of this book towards questions of the social legitimacy of transition. As Timothy Mitchell (2011: 238, 267) argues, in relation to renewables, 'the building of solutions to future energy needs is also the building of new forms of collective life', shaped by highly contingent 'battles over the shape of future energy systems'.

The USG has been a critical site of transition in the state, with over 1 GW of renewable generation capacity now built, and more on the way. However, the regional story detailed in this chapter raises many questions about 'energy transformation' whereby 'privately owned corporations … leverage private investment and public subsidies to fund the switch to renewable energy, effectively strengthening the capitalist status quo' (Cahill 2018: 23). What sort of transformation is this?

The conceptual depth and scope of social legitimacy encompass tensions and contradictions among the state, corporate, and civil society spheres. In this chapter, we have analysed these contradictions in renewable energy through the lens of USG host communities and civil society and documented the way in which the region has been shaped by a range of technological, identity, and locational concerns that arise from the process of appropriating wind and solar power. Additionally, we have pointed to related concerns about the distributive injustices of the developments, reflecting the overarching private sector-based model of accumulation.

We conclude that the social legitimacy of renewable energy development in the USG, based on ethnographic research focused on host communities and civil society, began strongly with RPA but is now weaker, despite some clear advantages for some groups. The processes of appropriation, accumulation, and regulation have been uneven across the region, with Port Augusta impacted early and most intensively, leaving a level of disenchantment and disengagement after the failure of the Solar Reserve CST project and the subsequent shortcomings of the renewables industry to fulfil the 'social contract' expectations. Despite this, and perhaps because of the regional learning that has taken place off the back of the Port Augusta experience, the forthcoming wave of renewables and hydrogen developments announced around Port Pirie and Whyalla continue to engender hope for socio-economic outcomes.

Some in the region have benefitted in various ways from renewables as we have documented. Host landowners and, recently, native title corporations are considerably advantaged with revenue flows and other forms of value. Native title groups in particular have created an opportunity for ground-breaking recognition of their rights to negotiate benefits and to share in the ownership of renewable energy. These forms of self-determination are likely to improve many aspects of well-being, outside the parameters of individualised market competition, potentially delivering a true change in the conditions of collective life, in Timothy Mitchell's terms.

Others in the region have experienced no change in their conditions of social and economic life, and still others have experienced some serious disbenefits such as psychological distress, dispossession of Aboriginal lands, dust-related health problems, unaffordable housing and homelessness, or a string of insecure work. The regulatory process and wider structures of energy governance reflected in regional planning too produce their own legitimacy challenges, especially in terms of procedural fairness and inclusion, and in terms of defining regional identity and interests. As we documented, residents' challenges and critiques of regulatory frameworks have at times won some improvements.

The legitimacy of renewable energy in the region is shaken by the fact that the regional gap is not closing and on some broad indicators may have been increasing over the years of the renewables growth that we studied in the USG. Importantly, we know there are not a large number of ongoing jobs in renewable energy generation during the 30 plus years of operational life. Most of the local jobs are during the construction phase and are short-term and lower-skilled. This is an important limitation of the energy transition as it is currently unfolding in the USG. Corporate renewables cannot address the acute and entrenched problems of regional unemployment, nor have renewables so far offered sustained opportunities in other segments of the local economy.

The USG journey towards the renewable energy 'benefits mirage' continues. Will it, or even can it, become a reality within the current paradigm? In public announcements, politicians now declare that the huge scale of hydrogen infrastructure and export investment that has been planned since we concluded our research will bring benefits to the region: 'We can use this clean hydrogen to firm our electricity grid, but more than that, we can use it to help reindustrialise the USG, creating thousands of jobs in the process' (Malinauskas 2023: 1). The mixed socio-economic results of renewables growth so far in the USG reveal the shortcomings of the dominant model of state incentives and privatised investment. It is imperative to expand the approach. 'Upstream' public policy, such as using 'wellbeing economy' measures and local transition authorities, could help keep these socio-economic objectives in focus. Facilitating more public and community ownership too could offer greater balance, as these models help to diversify the decision-making power and distribute the benefits. As our research in the USG demonstrates, a mindful new combination of approaches to renewable energy roll-out is required to maintain the brisk pace of carbon emissions reduction whilst also intentionally addressing endemic regional disadvantage and local host community aspirations.

Appendix A *Large-scale Renewable Energy Projects Upper Spencer Gulf, February 2024*

Name + (Initial DA Lodged)	Council Area	Initial DA Approval	Majority owner (February 2024)	Type	Stated capacity	Status (February 2024)
PORT AUGUSTA						
Lincoln Gap Wind Farm (2004 and 2019 [Stage 3])	Outback Authority (very close to Port Augusta)	2006 and 2020	Nexif Energy and RATCH	Wind + battery	468 MW wind (3 stages) + 130 MW battery	S1 operating 2020; S2 operating 2022; S3 pre-construction
Sundrop Farms (2009)	Port Augusta	Trial project 2009 (PA Council) 2014 (expansion)	Morrison and Co.	Concentrated solar thermal, and later additional solar PV	39 MW	Operational 2010/2017
Bungala Solar Farm (2015)	Port Augusta and Outback Authority	2016	Enel Green Power	Solar PV	275 MW (2 stages)	S1 part- operating 2018; S1 and S2 operating 2020
Port Augusta Renewable Energy Park (2015)	Port Augusta and Mount Remarkable	2016, 2019 revised	Iberdrola	Wind + solar PV	210 MW wind + 107 MW solar	Wind and solar operating October 2022
Aurora Solar Thermal Project (2017)	Outback Authority (very close to Port Augusta)	2018	Solar Reserve	Concentrated solar thermal	150 MW solar thermal	Cancelled 2019
Aurora Energy Project (existing DA)	Outback Authority (very close to Port Augusta)	2018 – purchased Aurora CST. DA approvals from Solar Reserve 2021 battery approval	1414 Degrees and Vast Solar	Solar PV + solar thermal + battery	150 MW solar thermal, 70 MW solar, 140 MW battery	Pre-construction
Playford Battery (2019)	Port Augusta	2020	SIMEC Energy Australia	Battery	100 MW	Pre-construction (no active progress

MOUNT REMARKABLE						
Baroota Pumped Hydro Project (2018)	Mount Remarkable	2019	UPC/AC Renewables	Pumped hydro + battery	250 MW	Pre-construction (no active progress)
Port Augusta Renewable Energy Park crosses into the District Council of Mount Remarkable. See project details in the Port Augusta section earlier.						
PORT PIRIE						
Clements Gap Wind Farm (2001)	Port Pirie	2004	Pacific Hydro	Wind	57 MW	Operational 2010
Crystal Brook Energy Park (2017)	Port Pirie	2019	Neoen	Wind + solar PV + battery	150 MW wind + 125 MW solar + 130 MW/400 MWh battery	Pre-construction (no active progress)
Bungama Solar Farm (2019)	Port Pirie	2019 solar farm and 2023 variation with battery	Amp Energy	Solar PV + battery	280 MW + 140 MW	Pre-construction
Clements Gap Battery (2020)	Port Pirie	2021	Pacific Hydro	Battery	60 MW/ 2 MWh	Pre-construction
WHYALLA						
Whyalla Solar Farm (2017)	Whyalla	2018	Adani	Solar PV	140 MW	Pre- construction (no active progress)
Cultana Solar Farm (2018)	Whyalla	2019	SIMEC Energy Australia	Solar PV	280 MW	Pre-construction
Yoorndoo Ilga Solar (2022)	Whyalla	2024 Project-specific Native Vegetation Act amendments approved 2023	EPS/Amp Energy	Solar PV + battery	300 MW solar, 250 MW battery	Pre-construction

References

ABC News (2012a). 'Port Augusta Backs Solar Thermal Power', ABC News, 23 July 2012.

ABC News (2012b). 'Minister on Solar Thermal Fact-Finding Tour', ABC News, 2 August 2012.

ABC News (2015). 'Alinta Energy to Close SA Power Stations, 440 Workers Affected', ABC News, 11 June 2015.

ABC News (2019). 'This Solar Plant Was Meant to Supply Energy to the SA Government for 20 Years. It's Been Scrapped', ABC News, 5 April 2019.

AIATSIS (n.d.). 'About PBCs', Canberra: Australian Institute of Aboriginal and Torres Strait Islander Studies.

Australian Bureau of Statistics (2021). *South Australia 2021 Census All Persons QuickStats*, Canberra: ABS.

Australian Bureau of Statistics (2022). 'More Growth in the Regions during the Pandemic', ABS Media Release, 29 March 2022.

Barngarla Aboriginal Determination Corporation (2022). 'Barngarla Win Special Commendation at the SA Premier's Energy Award', Media Release, December 2022.

BDAC and MAV Media SA (2021). 'Yoorndoo Ilga Solar: Barngarla Determination Aboriginal Corporation Video for Community Consultation', Barngarla Aboriginal Determination Corporation and MAV Media SA, 24 September 2021.

Becker, S., & Naumann, M. (2018). 'Energiekonflikte erkennen und nutzen'. In Kühne, O. & Weber, F., eds., *Bausteine der Energiewende*, Wiesbaden: Springer Fachmedien, pp. 509–522.

Beyond Zero Emissions (2012). *Repowering Port Augusta*. Port Augusta: BZE.

Bungala Aboriginal Corporation (n.d.). 'About Us'. Bungala Aboriginal Corporation. Accessed 15 January 2024.

Cahill, A. (2018). Economic diversity in the energy sector: post-capitalism in the here and now? *Australian Quarterly*, 89(2), 22–27.

Channel 10 (2012). 'Repower Port Augusta Rally for Solar', Channel 10 News.

Clarke, D. (2018). 'Why Did South Australia Council Reject Solar, Wind, Hydrogen Hub?' Renew Economy, 6 August 2018.

Combet, G. (2012). 'Securing a Clean Energy Future: Implementing the Australian Government's Climate Change Plan', Media Release, Australian Government Department of Climate Change and Energy Efficiency.

Commonwealth of Australia (1993). *Native Title Act 1993 (Cth)*, Canberra: Commonwealth of Australia.

Cullen, S. (2012). 'Govt Scraps Plans to Shut Dirty Power Stations', ABC News, 4 September 2012.

Cultana Industrial Estate Crown Land Transfer (2020). Agreements, Treaties, and Negotiated Settlements Project. 30 November 2020.

DP Energy (n.d.). *About – DP Energy*, Adelaide: DP Energy.

Federal Court of Australia (1998). *Nukunu People v State of South Australia*, Adelaide: FcoA.

Federal Court of Australia (2016). *Barngarla Native Title Claim Group v State of South Australia*, Adelaide: FCoA.

Federal Court of Australia (2021). *Barngarla Native Title Claim Group v State of South Australia, Port Augusta Proceedings*, Adelaide: FCoA.

Filatoff, N. (2020). 'SA Government Contracts for Energy to Advance Cultana Solar Farm and Playford Utility Battery', PV Magazine Australia, 2 November 2020.

Government of South Australia (2020). 'Across Government Electricity Retail Agreement', SA Tenders and Contracts, 30 October 2020.

Government of South Australia (2023). 'Hydrogen and Renewable Energy Act 2023', Government of South Australia, 7 December 2023.

Green, A. (2019). 'Energy Park Plans: Increased Turbine Height Causes Concern'. The Transcontinental, 19 June 2019.

Legatus Group (2018). *Rating Equity for Commercial and/or Industrial Land Uses Outside Towns in South Australia*, Adelaide: Local Government Association of South Australia.

Malinauskas, P. (2023). 'Preferred Partners Selected for Hydrogen Jobs Plan in Whyalla', Premier of South Australia Media Release, 22 October 2023.

Mitchell, T. (2011). *Carbon Democracy: Political Power in the Age of Oil*, London: Verso Books.

Moreton-Robinson, A. (2015). *The White Possessive: Property, Power, and Indigenous Sovereignty*, Minneapolis: University of Minnesota Press.

Neoen Australia (2018a). 'Crown Development for Crystal Brook Energy Park'. Development Number 354/V003/18, Neoen Australia.

Neoen Australia (2018b). 'Crystal Brook Energy Park: Development Application Response to Verbal Submissions', Neoen, 10 December 2018.

Office of the Registrar of Indigenous Corporations (n.d.). *About the CATSI Act*, Adelaide: Government of South Australia.

OpenNEM (2024). 'OpenNEM: South Australia', Media Release, 15 January 2024.

PlanSA (2023). 'Crown Development and Essential Infrastructure', Media Release PlanSA, 25 August 2023.

REVE (2020). 'The Closure of Solar Reserve, an Isolated Case of the Concentrated Solar Power Industry', REVE, 7 February 2020.

Roberts, G. & Gillian, A. (2022). 'After 28 Years, One of Australia's Longest-Running Native Title Claims Is Resolved in 10 Minutes', ABC News, 3 February 2022.

SA State Planning Commission (2019). *Discussion Paper on Proposed Changes to Renewable Energy Policy in the Planning and Design Code*, Adelaide: SPC.

Shteyman, J. (2023). 'Wind, Solar and Green Hydrogen to Save Whyalla from Coalition's Promised Wipe-Out', Renew Economy, 26 September 2023.

South Australia Department for Environment and Water (n.d.). *Pastoral Leases in South Australia*, Adelaide: Department for Environment and Water.

South Australia Pastoral Board (2016). *Pastoral Board Annual Report*, Adelaide: SAPB.

Spencer Gulf Cities (2022). 'Regional Partnership Priorities Launch', SGC Media Release, 13 September 2022.

Spencer Gulf Cities (n.d.). 'About the Upper Spencer Gulf Common Purpose Group', Port Pirie, n.d.

Starick, P. (2024). '$1bn empire: Aboriginal corp breaks wealth record in land deals', The Advertiser, 23 February 2024.

Stellar Health Solutions (n.d.). *Services*, Adelaide: Stellar Health Solutions.

Stephens, R., Cansdale, D. & Forbes, L. (2022). 'I haven't seen it like this before: calls to make soaring regional migration sustainable', ABC News, 18 February 2022.

Upper Spencer Gulf Jobs (n.d.). 'Upper Spencer Gulf Jobs', Port Augusta, USGJ.

Webb, L. (2019). *Assessment Report: Crystal Brook Energy Park*, Adelaide: State Commission Assessment Panel.

6

Comparative Analysis

The case studies outlined in this book demonstrate how the shift to renewable energy is structured by the socio-ecological relations of the energy transition. We have sought to do this across three dimensions – accumulation, appropriation, and regulation. There is contestation over ecological appropriation in terms of the scale and siting of installations and the technologies they use. There can be intense disputes over the spoils of transition in terms of the accumulation process and whether benefit should flow to corporates or to wider society. Structures of regulation are the focus for intense disputation in terms of demands for democratic participation and engagement of wider society in the transition process. This concluding chapter reflects on the three dimensions in light of the case studies and seeks to progress the analysis through a schema for interpreting transitions. The aim is to assist in anticipatory policy making in what is fast becoming a renewable mode of accumulation.

Socio-ecological Transitions in Large-Scale Renewable Energy

The process of energy transition is both temporal and spatial, mapped over time from fossil fuel reliance to renewable energy. The anticipation of a global energy transition, of what this may bring in terms of a new wave of accumulation, is central to understanding the spatial dynamics. The renewable energy dynamic is part of the wider process of both apprehending the climate crisis and seeking to address it through emissions reduction. Climate change imposes its own 'calendar', as Chakrabarty puts it (2022), for social agency and transformation beyond the dynamics of fossil capital. The overarching climate calendar can recruit a range of related and complementary calendars: there are overlapping time trajectories for local environmental sustainability, livelihood, energy affordability and access, industrial and economic development, health and community development, energy security, and, not least, the political processes mediating these issues.

These may be understood as the 'co-benefits' of a shift from fossil fuels, or more expansively, as part of the new socio-ecology of capitalism offered or promised by renewable energies. As these co-benefits are articulated into developmentalist frames, variously for a 'green economy' or a 'green new deal', they offer the possibility of a new mode of regulation. This creates a temporal horizon for ecological modernisation, whether defined as 'green' neoliberal capitalism or as a post-neoliberal capitalism or perhaps post-capitalism. That horizon is never fixed or stable: the current neoliberal transition creates both winners and losers, generating new contestations and forcing new agendas into view. Neoliberal accumulation generates a backlash not just from those excluded from the renewables goldrush but also from ecology itself as climate disruption continues. Climate crisis forces an urgency into the mix, creating cumulative pressure on national state authorities, threatening their political legitimacy, and creating new demands for democratic participation.

There are alternative pathways within the transition. The trajectory towards large-scale privately owned renewables is not preordained. The socio-ecological 'fix' that positioned thermal power stations powered by coal, gas, or oil as the chief means of generating electric power is being progressively unravelled in favour of a new 'fix' or 'fixes'. This is centred mainly on large-scale corporate and private investor-owned renewables, in some regions through a steadily increasing share of renewable energy, and in others through the wholesale replacement of fossil fuels by renewables. With and against this, there are a variety of models of household, aggregated, community, municipal, and state-owned renewable energy, on-grid and off-grid. While state agencies and corporate entities strive to 'lock in' monopoly power for large-scale renewable energy, capturing market share to secure rolling returns, the 'calendar' of renewable energy development and the related social organisation of energy has its own dynamics. As demonstrated by the rise of household solar, renewable energy has a distributive capacity that can sidestep attempts at 'fixing' supply and markets in monopoly terms. As renewables rely on broadly available sources of energy, they have the inherent capacity to develop a 'calendar' that is not easily predicted or controlled. Socio-technical certainty is only further disrupted by the anticipation of intensifying climate change effects and a requirement for 'de-growth' measures well beyond the existing policy imagination.

In these ways, the utility-scale renewables' fix' of today, and the regions that embrace it, may well become 'stranded' tomorrow. The energy transition increasingly rests on a particular accumulation strategy that may be vulnerable to changes in social, technological, climate, and ecological change. Counter-intuitively for the neoliberal frame, models of energy transition that do not hinge on a rush to capitalise and commodify renewable energy for profit, but instead are directed at municipal, community, and household-level adoption, may prove more resilient. Such models of local ownership can have an important spatial dimension of not rendering regions hostage to renewables capital and an emergent renewable-industrial complex. These aspects point to the process of energy transition as being inherently contested and open-ended, as an interim 'fix' subject to the ongoing exercise of social agency and socio-technical transformations under advancing climate change.

Appropriation

The process of securing renewable energy supply on the ground can be conceptualised as a process of capitalisation. Capitalisation allows an ecological attribute or process to be captured and transformed into an economic 'resource'. In this case our focus is on wind and solar power, though the same process applies for instance to transforming a mineral deposit into a mining resource (with the important difference that a mining resource is exhaustible). In all cases, the state approves, and legitimises appropriation, though there are also a range of non-state players that claim a role at different times and in different institutional contexts. The renewable energy 'prospector' is the first player, as they seek-out potential renewable energy sites and map for wind speeds and solar radiance. The ensuing transition process then draws together a widening range of participants, varying with modes of technology, institutional capacity, social power, market apparatuses, and structures of authority. Our aim has been to identify the participants, comparing across the three regional contexts to understand the three key dimensions: we argue these constitute capitalisation and shape the process of accumulation-by-appropriation through renewable energy production.

The critical conditioning factor for capitalisation is that socio-ecology is finite: ecology is not produced by accumulation but is necessary to it. Even relatively abundant sources of renewable energy are rendered finite by the bundling of factors necessary for profitable capitalisation, from land access to transmission infrastructure. Accumulation depends on appropriation but cannot control the availability of the resources that it requires. These resources are, as Polanyi emphasised, 'fictitious' in the sense that they are not produced by capitalism but are necessary to it (Polanyi 1944). As outlined in Chapter 1, abundance, scarcity, and exhaustion are uneven and impose a particular trajectory on regional development, for instance as a coal-producing or oil-producing region. For renewable energy, the 'resource' is uneven and its exploitation is shaped by its intersection with pre-existing development dynamics. Most important, the finite character of ecology generates recurring crises of exhaustion, that magnify and intensify over time, as they move from local to planetary scope (climate change being a case in point). It is this autonomous logic of ecology, its irreducible biophysical character, that drives capitalism's crises of exhaustion and the constant searches for the next 'fix'.

Within the cases analysed here, the logic of appropriation is reflected in the nexus between renewable energy technology, social organisation, and land use changes. As the technologies and policies change, so does the form of renewable development, producing different outcomes in terms of the exploitation of renewable 'resources' and in terms of land use change. The resulting uneven distribution of benefits fuels inequalities and contestation. We see this dynamic in play very directly with some classes of landowners benefitting while other landed classes, landless workers, and Indigenous peoples lose out. The pattern is clear across Karnataka, South Australia, and Brandenburg, where the unevenness of renewable energy development exacerbates inequalities principally along lines of land ownership.

The local legitimacy of the renewable energy transition hinges on struggles over the appropriation of land for wind and solar installations. In our cases, renewable energy is confronting very different patterns of land ownership, distribution, and use in the three contexts. In each case we find that the energy transition empowers and enriches landholders, and thereby both affirms and sharpens pre-existing patterns of land stratification. The question of unequal benefit becomes an object of policy and regulation, in the process politicising the land question in new ways. Contestation over the outcomes of renewable energy development can thereby create new agendas: questions of land justice and livelihood for smallholders and the landless may gain greater salience in Karnataka; affected rural communities and leasehold farmers may become more mobilised in Brandenburg; Indigenous land claims may make progress against the huge pastoral leases in rural South Australia.

Renewable energy is vulnerable to accusations it benefits only the already privileged, and that vulnerability can be leveraged to secure wider agendas. As documented in the case studies, increasing inequality and struggles over the appropriation of land and ecologies are putting pressure on developers and states to find alternatives to the extractivist model and to rebalance the new inequalities created by renewable energy. In Karnataka, this takes the form of proposals for mixed-use, agri-solar energy, allowing farming to co-exist with solar plants and requiring new co-management arrangements. In Brandenburg, revenue-sharing models have begun emerging to distribute income from renewable energy across landowners in zones of renewable energy development, rather than only to the landholders who directly host installations. In South Australia, Native Title corporations are beginning to access a significant share of the value of renewable enterprises through rents, share equity, jobs, and training for members.

Accumulation

As has been outlined, the globally dominant mode of energy transition centres on the process of capitalising ecology in terms of bankable or 'harvestable' wind and sun. To be incorporated onto the balance sheet of a company, whether state-owned or privately run, the biophysical force of wind and sun must be enclosed and secured. The company must hold rights to exploit the resource, or to sell the right to exploit, as theirs alone. The asset then is entered into the ledger, signifying the process of accumulation-by-appropriation. In this way, renewable energy, much like any other commodity, is constructed as an asset that extends across space and time, well beyond the boundaries of the solar panel or wind turbine installation.

Here, ecological value is subsumed into the money economy, accorded a dollar value that can rise and fall according to the business cycle and wider societal and economic trends. Importantly, an asset is valued in terms of its capacity to generate future profits: this is what makes it investible (Christophers 2024). With its future value measured against other assets, renewable energy becomes commensurable and fungible, integrated into global circuits of capital. Within financial contexts, renewable energy is not valued in itself, as 'use value', but in relation to other assets in terms of its value on the market, its relative exchange value. Finance is agnostic on the question of emissions and energy investors routinely seek to spread their risk, holding fossil fuel assets as a 'hedge' to offset any risks associated with renewables, and vice versa. For the prudential investor blind to climate concerns, investment in renewables then can simultaneously support fossil fuel investment.

The problem with renewables financialisation is that renewable energy is not 'just' a commodity. Planetary climate stability now hinges on the capacity to decarbonise energy use (or to reduce energy use). It stands at the centre of the common heritage of humankind, as having intrinsic value well beyond 'the market'. As outlined by IRENA (2023), the peak intergovernmental authority on renewable energy, investment in renewable energy installations now offers the greatest hope for climate stabilisation. Yet the vehicle for enabling transition – private sector finance – will not make the requisite investment without at the same time making a surplus. Capital is generally indifferent to climate change notwithstanding the across-the-board move to 'climate friendly' investment strategies (Bryant and Webber 2024). Renewable energy assets are still valued mainly in terms of future expectations of income flow under the global energy transition, not in terms of their contribution to emission reduction. Under this model, if renewable energy installations prove unprofitable, or even simply less profitable than other kinds of investment (such as in fossil fuels), they will not be constructed.

The only means of ensuring society is able to reach emission-reduction targets is through state action. This points to the issues of social and democratic mobilisation and state regulation addressed in the third dimension of our framework. Capital primarily responds to 'transition risk' in terms of the threats or opportunities of state-led climate policy, rather than to physical climate risks. As with any large-scale market-led transition, expectations are highly volatile and can drive speculation and investor euphoria, and at the same time produce uncertainty, instability, and sudden losses of investor 'confidence'. State policy underwriting private investment can become highly contested and produce pressure for more direct public ownership and investment. Over time there are multiple challenges to the legitimacy of the process, with multiple lessons learned. In a fast-moving technological context, laggards are regionally fragmented (Australia), frontrunners are overtaken (Germany), latecomers build on the shoulders of others (India). In all contexts, as demonstrated by our region-level cases, the transition poses new socio-political challenges and therefore new possibilities for social change, pointing well beyond the neoliberal models that to date have been put in place to drive transition.

Most directly in terms of accumulation, we document how the reliance on large-scale privately owned renewable energy inevitably raises distributional questions of who wins, who loses, and what co-benefits are created. As we outline in the three cases, there is very little 'social ownership' of these installations. More often they are viewed with suspicion, as a source of income for large corporates, rather than as a means of offering good jobs and delivering cheap energy or even as a means of addressing climate priorities. Across our cases, expectations of profitability underpin decisions of private investors to develop renewable energy projects. However, accumulation cannot proceed without legitimacy, at least not for very long.

Across each of our cases, the accumulation dynamics of renewable energy, and the surplus it generates, are heavily contested. There are different types of financial claims, from voluntary revenue sharing to royalty payments, local taxation and co-ownership, and new models for involvement and veto power in decision-making. Landowners renting land to a large-scale solar plant in Karnataka, for instance, made claims for co-ownership and in Shimoga village councils (*Panchayats*) sought direct roles in planning and developing wind power. In Brandenburg, a range of structures are in place for benefit sharing in wind power, and there are recent commitments on solar power. These emphasise distributed and municipal initiatives for unused space for solar panels and have a focus on local-level delivery and planning, including local revenues and re-skilling. In South Australia, there are on-going demands for local taxation and the possibility of a new phase of co-ownership in renewable energy development opening up with Indigenous landowners taking equity stakes in one of the largest solar plants in the state.

Regulation/Rivalry

National states encourage sub-national authorities to establish the means of implementing – and exceeding – national commitments for transition to renewable energy. National-sub-national relations, inter-regional, and intra-regional dynamics are reconfigured as the energy transition offers new openings for regional elites, energy capital, and local action. A rush to capture the emerging renewables economy is evident across the three case studies and is expressed in forms of public-private facilitation across infrastructures of generation, transmission, and distribution. Distinctly regional frameworks for the organisation of renewable energy, along with associated region-level development narratives, put into play regional legitimation frameworks for advancing renewables. Transition plans engage a range of social forces in new ways, with a variety of regional co-benefits articulated into wider regional narratives. Constituencies are mobilised for the transition agendas, from international investors to individual householders. People become actively engaged in the shifting and often contested transition agendas.

Renewable extractivism thereby creates its own territoriality, mapped in the emerging renewable energy regions. Large installations have their own distinct logic, with wind and solar farms often located far away from fossil fuel-based industries. 'Just transitions' agendas are explicitly developed for fossil-fuel dependent communities, but they also emerge, albeit implicitly, in programmes for co-benefits in emerging renewable-focused localities, such as via employment and allied industrial development, community funding, or local co-ownership. As we have demonstrated in the three cases, the process hinges on state power and on prevailing contexts for social legitimation. The state makes laws to transform ecology into a 'resource' for capital, and these mechanisms are embedded in and express structures of social legitimation. As we have demonstrated, the ostensibly economic process of capitalising renewable energy hinges on the socio-political process, which can vary dramatically across places, and over time.

There are many aspects that may come into play: the electoral appeal of cheaper electricity; a desire for a renewables energy policy; community involvement in energy planning and renewables development; cheaper, reliable, autonomous energy for regional economic development and export; state agencies seeking to access international development finance; local capital engaged with private investment flows and renewables financialisation; urban health benefits and possibilities for rural renewal; and measures to enable decarbonisation and diversification in fossil fuel-dependent communities. In the three cases, these and other aspects are reflected in the contrasting regulatory regimes and outcomes.

Across each of the three case studies, we see contestation between the winners and losers of decarbonisation mediated by the process of state legitimisation. The participation of state bodies has become critical for securing community consent for renewables projects. In Karnataka, state agencies act as brokers to enable the acquisition and leasing of land, establish power purchase contracts for generators, and offer a 'one stop shop' to facilitate private sector proposals. Planning approvals have become a focus for contestation, thereby democratising otherwise distanced bureaucratic processes. In Brandenburg, regional planning bodies have become crucial sites for claim-making by citizens seeking to capture a share of the renewable energy surplus or to shape its development pathway. This is also visible in South Australia where communities have actively mobilised through the planning process. The success of these actions in changing the social conditions for renewable energy development has put pressure on governments and developers to generate new models of renewable energy that can secure local legitimacy.

In this context, legitimacy may be defined more as a vehicle for change than as a barrier to it. If renewable energy is recast in a way that expresses public priorities, its local legitimacy can enable rather than obstruct its progress. The IRENA hints at this, stating that 'The more the energy transition can help solve … broad challenges (pervasive problems related to human welfare and security, deeply embedded inequalities) the more its popular acceptance and legitimacy will rise, provided also that community needs and interests are well represented and integrated into transition planning' (IRENA 2023: 1).

Legitimacy for Energy Transition?

The interactions between renewable accumulation, appropriation, and regulation are shaping socio-ecological relations and decarbonisation pathways. As we demonstrate in the three case studies, the outcomes are in turn shaped by conditions for legitimation. Contestation over large-scale private-sector renewables forces new issues, concerns, and possibilities onto the agenda. Efforts at legitimation are often undertaken in anticipation of possible disputes to contain or prevent any challenges before they arise, not simply 'after the event'. As state regulation anticipates and forestalls challenges, a form of 'reflexive' renewability emerges. Renewable capital, as noted, may be highly sensitive to contestation, and likewise take an approach to land acquisition, community benefit, and labour supply designed to accommodate concerns. Issues of transmission and pricing and debates about regional planning and regional futures all play a similar preventative or pre-emptive purpose. With this, a social planning horizon for renewables opens up, offering a space for participatory interventions in an otherwise privatised field of decision-making.

More abstractly, the process of contestation over the legitimacy of renewable energy creates a space for the socio-ecological imagination, of projecting the shape of arrangements in a climate-changed and decarbonised world. Challenges are posed across a range of dimensions though as noted they tend to converge around the issue of financial surplus and social benefit. The reliance on private capital as the key driver and beneficiary of capitalisation, as reflected in the dominant utility-scale approach, patterns this socialisation process. The issue of the public good and the extent to which it is served (or not) by the privatised transition process becomes central, and to varying degrees the state is forced to directly address this.

In Karnataka, the local state in partnership with federal authorities deliberately created a publicly owned corporation to establish the Pavagada solar project. It was this state agency that built trust with landowners and established the leasing arrangements, felt to be necessary to ensure the plant would not become a target for land reform agitation. This legitimating structure in Pavagada effectively displaced and institutionalised the land conflict, insulating private renewable energy investors: as a result most discontent over the land issue was focused on the state agency and its public responsibilities rather than on private players. There are parallel issues of social benefit in Brandenburg, where, as noted, the requirement to minimise negative impacts and to make a contribution to local provision translates into legislated requirements. In South Australia's Upper Spencer Gulf, much contestation converges on the issue of regional benefit, where it is the failure of the state to establish a means of ensuring local benefit that is most questioned, rather than the corporates themselves. Again, it is the state that acts as arbiter and as facilitator in determining what is or is not considered to be a 'fair' outcome for local residents, insulating private capital from claims for instance that renewable energy should be publicly or community owned.

In this way, the transition process politicises energy in new ways and forces state institutions to address demands that they act as guarantor for the public good. Renewable energy thus socialises the climate crisis and decarbonisation into the structure of authority as expressed in state policy, including for instance the emergence of new 'special purpose' state agencies. This process of renewables development may appear to follow patterns common to any industrial sector – that is, decarbonisation is normalised as re-industrialisation. Ideologically, renewable energy may be effectively de-climatised and 'interpellated' into the existing *modus operandi*, simply as the latest of economic and technological development. Certainly, there is considerable overlap between the planning bodies, construction contractors, and operating companies in renewable energy and other industries. Companies accustomed to establishing and running large-scale energy infrastructure, including coal mines and coal-fired power stations, embark on renewable energy projects replicating many of the long-established assumptions about the framework for energy supply. Across all three case studies, there are specialist renewable energy companies, as well as corporations and investment funds whose portfolios stretch from renewable energy to the coal, oil, and gas industries and other industrial sectors such as transport and telecommunications.

At the same time, the broader climate goals of renewable energy can make the renewables re-industrialisation process uniquely politicised and disrupted. In this sense, the socialisation of decarbonisation sees development being forced into a new nexus with ecology, under a new climate-changed socio-ecological dispensation. As we have noted, policymakers and renewable energy companies can become anxious to maintain their 'green' credentials and become especially sensitive to criticism from governments, shareholders, and civil society. They may, indeed, be put in a difficult situation as the growth model does not deliver the emissions reductions that are required. Regardless, climate change is set to intensify, despite the success or otherwise of decarbonisation. The 'green' industrialisation model is put under a new form of socio-ecological pressure – it depends on 'green' credentials and will always be put into question for its inadequacy and be uniquely vulnerable to disruption by the biophysical forces it ostensibly is designed to address.

In this context, sharp tensions can emerge between the public interest in emissions reduction and the primary interests of private renewable energy developers. Contestation over capital's profit-take and the extent to which it can legitimately appropriate ecology's 'gifts' is central to the dynamic of regulation and contestation, as is the potential for counteracting tendencies, such as the emergence of a state-run, municipal, community, or householder-based renewables sector. This contest over surplus is effectively a form of displaced class struggle and is directly expressed in conflicts over labour hire practices and debates over regional futures. In this way, the questions posed by the energy transition open a series of fronts that can coalesce and become magnified in demands for public participation and public or community ownership in the newly capitalised ecology. These demands may pose a serious challenge to existing political elites and their capacity to manage local legitimation of the emerging sector, especially if it fails to deliver on its 'green' credentials.

Across the cases discussed in this book, we trace how contestation is reflected in the societal and political formations of renewables transitions. The case studies provide a wealth of data on the responses (and non-responses) of political formations as they relate to renewable energy. There may be local campaigns for or against renewables, or for specific social benefits and outcomes. Political parties and formal political representation may play a role, whether in advancing social participation or in blocking it. Renewable energy can become contested political territory, a site for contending political visions and alternatives and for wider engagement as an 'energy democracy'.

Whatever the normative status of these political schisms, they enter into the legitimisation process of renewable energy and shape the distinct region-level decarbonisation pathways. All these aspects are evident in the three cases. In Brandenburg, for instance, there have been concerns that the far-right 'Alternative for Germany' would exploit discontent at the impacts of renewables. In Karnataka, there are deep divisions between the BJP government, which narrowly won the 2018 election, and the Congress Party that regained office in 2023. These divisions translate into political rivalry over the impacts of renewable energy and 'just transitions' debates and locally between different villages dominated by either party. In Australia, renewable energy became a political scapegoat for the conservative federal government in 2013, with South Australia's Labor government characterised as having backed renewables at the expense of its people. That political dispute resolved when the opposition won political office in South Australia and adopted the previous government's renewable energy policies.

Renewables Contestation – A Schema

The three case studies discuss key points of contestation in the process of renewable energy development. There are parallel aspects, and here we explore the possibility for a comparative schema for analysis. We outline five broad aspects. These are each understood as critical components in the 'field' of renewable energy capitalisation. The spatio-temporal context is radically varied and is critical: the decarbonisation process occurs in various geographical spaces and institutional sites depending on the stage of renewable energy development at both the installation and regional scale. Yet, as we argue, there are common attributes allowing a shared conceptualisation and analysis, giving scope for common lessons to guide future development.

The schema partially disaggregates our three-part model of appropriation, accumulation, and regulation. In this we replicate and adapt Becker and Naumann's categories (2018), grounding them conceptually in the capitalisation process. We focus on five aspects of legitimacy: technological, locational, distributional, procedural, and developmental. In broad terms, these reflect the various roles and interactions of technology, the state, corporations, finance, land, labour, civil society, infrastructure, and consumption in shaping the renewable energy transition process. The five aspects are outlined in Table 6.1 along with a list of some related themes drawn from the case studies. This model builds on and elaborates the schema outlined in Chapter 1, as summarised in Table 1.1. They are discussed in more detail in what follows.

(i) Technology-Ecology Nexus

Renewable energy establishes a particular technology-ecology nexus as the bio-physical capacity to extract energy from ecology, mainly wind and solar. The biophysical conditions of appropriation are centrally important and set the parameters for many subsequent dimensions of contestation and conflict. Reliance on fossil fuel energy sources produces its own biophysical constraints and impacts, and likewise for renewable energy. While fossil fuel-based energy infrastructure is centred on sites of extraction and burning – coal mines, gas and oil fields, power

Table 6.1 *Conceptualising renewable energy capitalisation*

Renewable legitimacy	Underlying process	Themes from cases
1. Technological – technology-ecology nexus	Appropriation	Extracting a 'resource', innovation, type and size, rivalry with household, community, and state-owned models
2. Locational – land use, labour and community	Appropriation	Land contests, dispossession, stratification, competing uses, community dis/benefit
3. Distributional – capital flows and dynamics	Accumulation	Enclosures and surplus, rivalrous ownership, PPAs, inequality, livelihood and jobs, social ownership
4. Procedural – Regulation and re-regulation	Regulation	Planning and policy institutions, sub-national initiatives, regional rivalry, corporate capture, democratic demands
5. Developmental – regional future and identity	Regulation	Changed regional economy, 'downstream' uses, societal electrification, regional development strategy

stations, and ports – renewable energy can be located much more widely and much more visibly. It can be located far away from existing fossil fuel infrastructure, on 'greenfield' sites where the 'resource' may be strongest, and where land may be most easily accessed. Access to transmission grids is another important condition. Together, these aspects combine to impose a particular renewable energy geography.

There are substantial differences in the conditions for renewable sources: wind power, for instance, can be compatible with other land uses and may be more easily located on high ground not used for agriculture. However, in Brandenburg in particular, the increasing height of wind turbines and proximity to residential regions have become a focal point of technological conflict. Solar power generally requires a flatter, lower landscape, rather than high ground, and may come into conflict with existing land uses. In Karnataka, solar parks and their associated transition infrastructure have directly displaced existing agricultural industry and livelihood.

Importantly, these conditions change over time: innovation can act as a 'disruptor' that responds to the possibilities and problems of the existing technology. One example is the emergence of agri-solar, which is seen as allowing coexistence with agriculture, one of the key problems for solar power in Brandenburg, for instance. Utility-scale operations can themselves be undermined by innovation in distributed renewables. Rapid innovation to enable mass public access to renewable energy has seen extensive innovations in solar panels and also in small-scale storage, which in some contexts has siphoned off energy demand, threatening the business model of large-scale renewable energy installations. Policies across the three cases prevaricate between efforts at dampening householder rooftop renewables and harnessing them via aggregators, creating conflicting incentives and disincentives for investors and householders.

The different forms of renewable energy also have different potential and impact. Unlike fossil fuel electricity, which has mainly been centralised into large-scale utilities, renewable energy can be socially distributed, across households and communities. Grid access remains central, although the grid itself can be distributed, with various mechanisms for disaggregated small-scale renewable power to 'feed-in'. There is also pressure for localisation as transmission over large distances leads to energy loss. Whether locally generated or not, falling energy costs encourage increased consumption to drive profit and more generation. This can expand the scale of the model exponentially: South Australia is a case in point, moving from generating less than 1% of electricity from renewable energy in 2003 to 70% in 2023 and a target of 100% by 2030 (Department of Energy and Mining n.d.). The state now has an aspirational target of 500% renewable energy by 2050, linked to expanded electrification, green hydrogen and steel production, and strong growth in national and international exports (Parkinson 2023).

When established, renewable energy does not require continual input of non-renewable resources to generate electricity, and as such has a smaller ecological footprint. Renewable energy can complement coal-fired power initially, although the more it reduces the cost of daytime power, the more it undermines the viability of coal-fired thermal power stations. Thermal power stations have to operate continually and cannot be switched off during the day, meaning they become lossmaking unless compensated, for instance, via ongoing power-purchase agreements. This tension is at the centre of Karnataka's energy system, with daytime energy demand effectively reserved for thermal plants. Meanwhile, the renewable sector is envisaging and developing storage solutions, including for instance via 'green' hydrogen or ammonia, pumped hydro, and new types of batteries, including new possibilities in 'virtual' aggregated storage. This dynamic reflects extensive technological capacity in the renewables sector, where there is wide scope for innovation to address blockages and reframe possibilities, well beyond the existing models.

Renewable energy landscapes are produced at this technology-ecology nexus and set the frame for much of the contested politics of decarbonisation. In Brandenburg, the main source of renewable energy is wind power, reflecting the early focus on community and municipal wind development along with political and economic incentives. Changing solar technology and the reduced availability of sites for onshore wind power are now enabling the growth in solar power.

In South Australia, the focus on utility-scale renewables has also been on wind power, primarily to capture high coastal winds and to access large tracts of upland pastureland; in contrast, solar power has been mainly household-based, although that is changing with new large-scale solar utilities. In Karnataka both solar and wind are well developed, with a stronger recent emphasis on solar power through large state-supported utilities, reflecting the global trend towards the acceleration of solar compared with wind (IRENA 2023). State facilitation and economies of scale for corporate players are more easily achieved at these large solar plants. Local ecology and the existing industrial geography, especially in terms of transmission, are important factors, along with technology. State policy mediates across these aspects, as discussed later.

(ii) Land Use, Labour, and Community

There is no renewable energy without access to land (or alternatively to a seabed, rooftop, or water surface). Social and biophysical limits to land are the main factors that define wind or solar power as limited 'resources'. Land may be widely available, for instance in South Australia, but there are particular requirements for renewable energy, depending on the technology. There is only so much land available in the localities that have strong winds or high solar radiation that also have access to the electricity transmission grid and have a ready labour force able to staff the operations. Here the pattern of settlement and land tenure comes to the fore. Renewables investors seek an empty space for accumulation, exemplified in the sea of solar panels in the Pavagada solar park. In doing so, they operationalise existing land inequalities, from Australia's myth of 'terra nullius' to India's failures in land reform.

In some contexts, such as in India, the capacity to gain access to land for renewables can be the overriding consideration. Land is linked very closely with livelihood in India, and especially so in rural areas: to be landless is to be potentially destitute; small-scale landholdings represent a critical foundation of social security and food supply. In these contexts, the state may be anxious to ensure renewable energy installations benefit local landholders, although by the same token, as we found in Karnataka, this excludes local landless populations or small-holders who may rely on agricultural work and on access to common land. In contrast, in Germany, land may be more commonly valued as a living environment and as a basis for community life. Here the renewables sector is called upon to respect the wishes of the local community, who may value the landscape as it is and who may not wish to live next door to large wind turbines with no discernible local benefits. In Australia, properties in rural areas may be owned freehold but very often are farmed as long-term pastoral leases on government or 'Crown' land. As discussed in Chapter 5, these pasture leases, and certain other forms of government land, can be subject to Native Title claims by Indigenous Traditional Owners. Change of land use must be negotiated with owners, leaseholders, and Native Title groups or claimants, with appropriate forms of rent and other compensation.

Across all three cases, the priority and precondition for establishing renewable energy installations is to gain landholder assent. In the process, substantial financial benefits may be committed to cover leasing and acquisition arrangements. Those who do not own the land, or whose land is deemed not suitable for renewable energy development, are generally excluded from benefit. Renewable energy projects can thereby exacerbate pre-existing social divisions between those who own large amounts of land suitable for renewables, those who own land that is not used, and those who do not own land at all. In India, the division between landed and landless is especially sharp: renewable projects can literally remove the livelihood of landless people insofar as they depend on agricultural labour or access to common pastureland. Parallel issues emerge in South Australia, where wind power may be located on large tracts of pastoral leasehold land. Renewable energy thereby benefits landholders descended from the original recipients of land grants from the colonial administration (which Aboriginal people have occupied for many thousands of years). Other residents without landholdings in the project area may miss out on benefits, though they may gain from short-term jobs and commercial activity in host towns, especially during construction phases. In Germany, there may be tensions between farmers who benefit from access agreements and other residents; with solar power there is the potential that leaseholders, who make up the bulk of farmers, will be displaced as landowners convert to much more profitable solar power. The lesson overall is that those without land are excluded from direct benefit, increasing inequalities.

Some of these impacts can be addressed through other forms of community benefit that are more collective in character, in the form of tax revenue, such as in Karnataka, where renewable energy companies are required to contribute to a CSR community fund; there may be revenue for local authorities or even direct payments for residents affected by installations, as have been instituted in Brandenburg. There is no legislated requirement for renewable energy companies to make such payments in South Australia, either to contribute to local councils or to provide community funding, though ad hoc voluntary contributions do occur.

The question of community benefit is overlaid with the issue of labour supply and employment. Projected job creation is central to regulatory approval and to the social legitimacy of renewable energy development plans. But the distribution and quality of these jobs, the extent to which they are available to locals, and even the definition of what counts as a renewable energy job, are heavily contested. In practice, there are often fewer jobs available than are promised and the available work tends to be 'frontloaded' in the construction and development phase rather than in on-going operations and maintenance, which generally require fewer workers. Labour issues were acute in Karnataka and South Australia, where the case for renewable energy is more embedded in local and regional development strategies and projections. Promises are made to the local community via the development planning process, and the failure to deliver on these promises can become a key source of disenchantment and antagonism.

Utility-scale renewables development can clearly accentuate regional disadvantage and access to jobs can be one of the most salient everyday expressions of this. Jobs are often allocated to people from outside the locality in order to access required skills, while local jobs are less skilled. Construction contractors may prefer not to employ from the locality, training their own mobile crews and maintaining a stable workforce. If they do employ locals, they will often do so through labour hire companies rather than directly, signalling the complex and multi-layered set of contracting and subcontracting that shapes the renewable energy workforce. Unlike the state-owned fossil fuel sector in India, jobs in renewables are private sector jobs often with minimal standards. Employment relations are often precarious, and unionisation is minimal. Companies can exploit the weak bargaining position of local workers who may be located in areas with high unemployment, as in Upper Spencer Gulf, or are in drought-ridden arid zones, such as in Karnataka.

In South Australia and also in Karnataka, many of the more specialised and higher-paid jobs, especially in maintenance, are carried out by workers who move between different renewable energy regions. In Karnataka, there was considerable resentment from local landowners who had expected skilled work, but only had access to work as security guards, or for solar panel cleaning and weeding. Landless people had to compete with higher-status landowners for the same pool of unskilled work and had even less access to paid work. The failure to meet local expectations generates demands for more extensive and proactive measures. The lack of local skills raises the question of why training for local workers cannot be offered and poses the need for more active government provision, in particular to integrate education, training, and skills programmes within renewable energy policy frameworks. This was expressed in demands for local training programmes in Karnataka, and also in South Australia. In some contexts, the local workforce may have leverage, especially if they have been displaced from fossil fuel operations and the government has instituted retraining and job creation programmes as part of the transition, as is anticipated in Brandenburg.

Conflicts over land, labour, and community benefit are usually not directly concerned with the legitimacy of renewable energy in itself but rather the environmental and social conditions that make it possible and its broader socio-economic effects. These concerns can be addressed in advance through regulation, incorporating community benefit, land justice, and decent work into the process of establishing solar and wind power. Local people in general argued that utility-scale wind and solar plants should be treated as part of the wider regional development project, not simply as power projects. Calls for development concerns to be uppermost were heard in the many phases of project development in each of our study localities, from those most marginalised to those most likely to benefit.

(iii) Capital Flows and Dynamics

Flows of investment capital for renewable energy have played an important role in shaping transition outcomes, and not least in the three country cases examined here. All three countries have undergone a process of neoliberalisation in energy supply, intensified from the 1990s with a process of electricity 'unbundling' across generation, distribution, and retail. Privatisation has proceeded opportunistically, along the lines of least resistance and highest returns: power generation has been one of the most profitable and easiest to privatise. Germany liberalised the energy market in the mid-1990s, which only intensified the oligopoly of generators and distributors, which was later slightly weakened by unbundling generation and distribution. Generation and retail were privatised across a number of states in Australia in the 1990s, with distribution ('poles and wires') and transmission later privatised unevenly across regions. India permitted private generation from the early 1990s, initially to supply 'captive' power for particular industrial uses and then more generally; efforts at privatising loss-making retail and distribution failed due to public pressure against price hikes. In India and Australia, the remaining state-owned aspects of the electricity system have been corporatised, while Germany's move to neoliberalism resulted in an increase of international investment as well as a (re)municipalisation trend in regional transmission and production. Aside from legacy hydropower, ownership of 'modern' renewable energy generation assets has largely been private across all contexts. This represents a different form of privatisation from one which involves selling off existing fossil fuel assets. The result is nonetheless an increasing share of private capital and ownership in electricity systems. The resulting public-private mix is overseen by a range of statutory agencies, at one remove from politicians and energy departments (though accountable to them via politically set regulatory frameworks).

The most important result from electricity privatisation has been the creation of privatised monopoly power, concentrated in the hands of a small number of large corporates. These corporates have the capacity to pursue their special interests through government, whether as fossil fuel companies or renewables companies, at the cost of wider public interests. This can make it very difficult for governments to undertake any policy initiative that 'harms' existing players. The articulation of private corporate interests – be they international or mainly national companies – with state electricity policy has become an important aspect of energy politics in each of the three regions. Renewable energy companies have their own priorities and especially in India and Australia often operate at a transnational scale far removed from the regional level. They may be headquartered on the other side of the world and establish country-level divisions to seek out 'prospective' sites and to influence potential regulatory outcomes to favour their interests. Increasingly, renewable energy installations are divided into individual plots or turbines, and owned as discrete, 'alternative' assets by investment funds (Christophers 2024). We have documented a range of these mechanisms across the three cases, showing how they can affect outcomes on the ground.

Public concern at the political power exerted by private electricity utilities has led some state authorities in Germany, at the municipal level, to (re-)nationalise local energy systems. Most prominent examples are the cities of Hamburg and Berlin, but between 2007 and 2012 alone, 60 municipal utilities were newly established and 170 regional grids (re-)communalised (Kungl 2018). Reflecting this, private renewable energy companies rely on state regulation and wider public legitimacy to progress their projects and can be highly sensitive to local pressures. They are competing with other companies in the sector and may have a degree of reputational capital to defend, especially where they are headquartered.

As part of an emerging sector, renewables companies may find it necessary to borrow funds to invest in large-scale renewables projects. This is not so important for the larger corporations with assets that may stretch across the energy sector and beyond, and hence are able to hedge against any difficulties that they may face in their renewables portfolios. Such companies may also be heavily invested in fossil fuel energy, or across other sectors of the economy, as is the case with the industrial conglomerates that have a large presence in the Indian renewable energy landscape. The Adani Group, for example, is one of the largest developers and operators of renewable energy in India, while also operating across coal and gas mining and power generation, port and airport, and other infrastructure sectors. Asset development can be structured as 'project finance', where an installation is constituted as its own private entity, as a special purpose vehicle. These are often created with joint ownership from different public and private stakeholders, on a 'partnership' model. Initiatives of this sort became commonplace in the 1990s to deliver public infrastructure, from toll roads to ports, across both high- and low-income countries. Adopting them for renewables, as has been common in Karnataka, for instance, simply transposes the delivery mode from one type of infrastructure to another. The type of financial vehicle that is used makes a difference on the ground as it shapes the relationship between renewable energy assets and the equity and debt markets, and taxpayer funding, that make renewable energy capitalisation possible. It also shapes the options, or non-options, for public input and participation.

Dynamics of both corporate concentration and de-concentration are visible and constantly shifting, and again have direct impacts. In the Pavagada solar park example, the 200 MW solar lot originally awarded to and developed by Japanese company SoftBank was transferred to Adani, which had already developed 150 MW at Pavagada, after it acquired SB Energy in 2021. Renewable energy may be used, for instance, to diversify investment by sector or technology, or to create separate specialised renewable energy companies to take advantage of the privileged position of renewables in the eyes of investors. Contracts between renewable energy producers and users are also closely related to the corporate strategies of large energy users from 'heavy' industry, such as steel and aluminium production to technology sectors, that are seeking to power their servers with green energy.

The construction of renewable energy as an asset also comes with a speculative dimension. Renewable energy-generating companies may be seeking a foothold in the regional energy market and be willing to engage in loss leadership to gain access. Bidding below the average price is essentially a bet on the likelihood that technology costs, and cost per unit, will fall still further. Like new tech industries, which are also rapidly changing and therefore subject to speculative investing, publicly traded renewable energy companies are at times heavily overvalued by investors with the value of their market capitalisation exceeding the value of income flows. As a tradeable asset, renewable energy stocks rise and fall with wider economic volatility. Renewable energy stocks have experienced periods of rapid inflation and deflation, and thus also high volatility, including an earlier round of boom and bust before and during the 2008 global financial crisis, and again during and following the COVID-19 pandemic. State authorities may be able to leverage investor euphoria to extract social benefit, but this can be a fragile basis on which to create sustained benefits from decarbonisation. That leverage is inherently insecure, not least as investors can shop around for other less burdensome renewables policy regimes.

There are wider pressures as the sector grows more crowded, with numerous energy corporates, including fossil fuel companies, now claiming 'green' credentials through their investment in renewables. Investors routinely screen companies for 'transition risk' (risk from climate policy), which itself encourages a flow of new funds into the sector (Goodman and Anderson 2021). Diversified investors may seek a reputational 'haven' in renewable energy and be concerned about any accusations of corporate wrongdoing. Renewable energy represents a highly attractive investment in an era of climate risk disclosure and environmental, social, and governance (ESG) ratings. As noted, most private sector 'climate finance' is directed towards renewables; as renewable energy is capitalised, it is being rendered as a standardised asset class with the predictable yields sought-after by large institutional investors. In practice, though, renewable energy technology is in constant flux, its social organisation is highly contested and fluid, and as a sector it remains highly volatile, putting a premium on the search for security of profit flow under the wing of the state (for instance via various offtake deals).

The sector organises its own business peaks and lobbying strategies, with renewables associations active across state regions, to national and global contexts. In all three national states, there are active renewables lobby organisations. These include the Clean Energy Council in Australia, the Indian Renewable Energy Federation, and the German Renewable Energy Federation (Bundesverband Erneuerbare Energie (BEE)). These lobbyists are counter-posed against the established fossil fuel and mining lobbies (coal in Brandenburg and Karnataka, and gas in South Australia). These assert their own interests and defend established state subsidies. Measuring the scope and type of private involvement in a comparative context helps reveal the extent to which companies shape outcomes in the contrasting contexts.

(iv) Regulation and Re-regulation

State institutions and their policies shape the possibilities for energy in general and renewable energy in particular, including the extent to which it is utility-focused, or indeed whether it is primarily state-owned or privatised. These policy dynamics are political and contested. In the case studies, we track these dynamics on the ground and how they shape outcomes at the regional level. Sub-national authorities have explicitly set targets for decarbonised electricity. In the three cases studied here, the region-level target for renewable energy substantially exceeds the national target and is used to drive policy formation at the region level. Policy success is then measured by the extent of renewable energy development.

Notwithstanding privatisation, renewable energy installations are heavily regulated by state authorities. Regulators determine key aspects such as transmission pricing and grid priority that directly shape possible returns for investors. In this context, policy is often geared to attracting private capital to invest in the generation of renewable energy, at the installation level. There are extensive tax incentives, public subsidies, and regulation for energy infrastructure and land acquisition in each of the three regions. There are also public–private partnerships offering price guarantees under power purchase agreements and related instruments, such as ‘contracts for difference’. Within the scope of such initiatives, there may be regulatory measures to ensure local benefit, but it is often unclear what is incorporated due to commercial secrecy requirements. Their material impact depends on the extent to which such non-price factors are prioritised in tendering decisions and enforced in practice. In the Karnataka case, price has reigned supreme in the tendering of power purchase agreements, driven by the needs of the state-owned Discoms in securing low-cost power, with social considerations relegated to other regulations and voluntary corporate contributions.

There is extensive evidence of the state taking a 'catalytic', indeed experimental role in fostering renewable transition, with a strong element of trial and error in policy innovation. Much of this is within the neoliberal frame, yet the general lesson is that neoliberal marketisation is always state-facilitated. Public finance underpins much of the private development of large-scale renewable energy. The development of the sector – and the capitalisation on which it depends – does not proceed without the active involvement of the state agencies. This dynamic politicises the process as public authority and public resources are directly orientated to private interests.

Private renewables depend on state regulation – and by the same token, state regulators can become increasingly dependent and even 'captured' by private renewable energy interests. Reflecting this, most officially measured 'climate finance' from public sources is now geared to facilitating privately run renewable energy (Buchner et al. 2023). Direct public investment and ownership remain common in transmission and distribution infrastructure, but in generation, the state is more focused on using its authority to 'de-risk' renewable energy to enable private investment (Gabor 2023). State-backed income streams enable renewable energy developers to access the credit they need to proceed at lower rates of interest. This provides states with considerable leverage over the shape of renewable energy development, though it is often unclear how (or if) this leverage is being used.

The dependence of renewable capital on state policies points to the possibilities for wider publics, including climate activists, unions, or local communities, to shape the terms on which public finance for renewables is extended. The increasing role of competitive tendering mechanisms such as reverse auctions, which have been used across all three of the case studies, means that price-based competition dominates, forcing other goals to the margins. Key issues for social legitimacy, such as sustained employment, regional development, or environmental impact, can be relegated to a lower order of priority.

The neoliberal model of tendering out renewable energy via power supply contracts also undercuts smaller-scale suppliers, leaving the field open for only the largest corporates and investors. It is instructive how the locally owned wind power has been forced out of the booming wind power sector in Karnataka; large-scale solar power is now almost exclusively controlled by the large corporate players. There has been a similar development in Germany, where contractualisation has marginalised cooperative, community, and municipal wind power, which had emerged in the early 1990s on the basis of feed-in tariff regimes. Reverse auctions were in part designed to make renewable energy competitive vis-à-vis fossil fuels, but in practice they can negate the wider benefits of transition, enable corporate concentration in the sector, and produce a 'race to the bottom' between renewable energy developers, technologies, and regions.

As forms of public finance, though, power contracts remain open to political contestation and to demands that social priorities should prevail over narrow pricing concerns. We find this politicisation of neoliberal renewables across the case study regions, where early promises do not translate into development opportunities. There are many aspects to this: as discussed earlier, the absence of decent jobs in the sector is a major issue in Karnataka and South Australia; the failure to involve local communities in planning and benefit-sharing is a key flashpoint in Brandenburg. Subnational authorities, citing the competitive regional investment environment, argue that requirements for developers to enter into commitments to deliver on social and environmental benefits pose a threat to corporate profit. In South Australia, some renewable energy companies have offered voluntary community grants as a model for delivering benefits as an alternative to state-mandated contributions. Community disappointment with this minimalist approach can become a significant barrier to social legitimacy for the sector.

Beyond this conundrum, the ownership structure of the sector can be centrally important in framing future possibilities for policy. Challenges to 'path dependency' can be critical, in terms of the extent to which each type of renewable energy becomes more concentrated in large-scale utilities or is more distributed. Regulatory models for encouraging households and communities to develop their own capacity for renewable energy may be in place, or regulations may actively discourage distributed provision. Outcomes vary substantially across cases and across types of renewable energy: in South Australia there is a proliferation of rooftop solar photovoltaics; in Brandenburg there is established medium-scale provision in wind power, encouraged by some state regulations; in Karnataka there are many small-scale wind generators and some larger-scale solar installations. Scale and type vary, as does unevenness. Renewable energy is spatially concentrated in varying ways. Large-scale solar may be in localities that have limited agricultural prospects, as is the case in Karnataka. 'Renewable energy zones' or 'hubs' may be designated to enable concentrated renewables development, such as in the Upper Spencer Gulf in South Australia. There may be strict regulatory limitations on where installations may be located, to protect existing forms of land use, such as in Germany. These planning decisions may have a deliberate 'development' purpose, as part of a local, subnational, or national strategy and be designed to enable new linkages within an emerging renewable economy (see v. later). This is evident in the South Australia case at the state-region and interstate level, and in Karnataka at the local level at the Pavagada solar park; in Brandenburg there is a greater emphasis on the impacts of renewable wind energy as a burden, to be delimited.

(v) Regional Future and Identity

The creation of designated renewable energy zones by planning and development authorities is often motivated by a desire to concentrate renewable energy generation – mainly to minimise the need for new grid infrastructure. In India, renewable energy relies on co-location with existing high-power transmission lines. The Pavagada power plant, for instance, links with India's main North-South transmission lines to the east of Karnataka. Grid access can reflect the prior industrial geography of fossil fuels, which can then be replicated in renewable energy. Parts of Brandenburg have a history of coal mining and coal-fired power, with high-powered transmission lines already in place. In Upper Spencer Gulf, the closure of the power station freed up transmission capacity, attracting renewables developers to the low-cost grid infrastructure for projects.

The concentration of large-scale renewables, whether in designated 'greenfield' zones or in pre-existing industrial areas, leads to a dynamic of regional agglomeration. Infrastructures, contractor networks, administrative supports, labour flows, and many other aspects of the locality become orientated to and converge on the emerging large-scale renewable industrial development. This reconfigures the region from an agricultural or industrialised locality into a renewables region, creating new development prospects. Regional authorities seek to accentuate these prospects, maximising opportunities for the locality both for existing residents and for newcomers. An intent to reverse the economic decline of rural areas can be an objective of regional planning bodies, as has been the case in Karnataka. Similarly for vulnerable industrial areas, renewable energy can be positioned as offering means of revitalising regional centres. In the Upper Spencer Gulf, the campaign for revitalisation and new jobs centred on the solar thermal project was a community-led initiative.

The change in regional identity that this brings can become a source of conflict and contestation. Agriculture is certainly affected by renewable energy installations. There can be major concerns about lost farmland and the resultant impact on farm workers, which was a major issue in Pavagada in Karnataka and is a pending issue in Brandenburg in terms of the displacement of leasehold farmers by in-coming solar installations. In South Australia's huge and sparsely populated northern land area, these concerns are not so strongly voiced. There also may be wider concerns about impacts for local ecology and landscape: this was apparent in Brandenburg and in some Upper Spencer Gulf projects, though with differences reflecting connections between identity, region, and land.

Renewable energy much more directly complements heavy industry as a source of cheap electricity and can be seen as enabling the reorientation and survival of on-going industrial activity, as well as allowing the emergence of new correlated 'downstream' industrial developments. The concerns here centre on the extent to which the emerging renewables sector can be required to enable this industrial diversification. This is a significant theme in the Upper Spencer Gulf region of South Australia, where the promise of renewable energy and the bulk of its proposed benefits for regional development are linked to visions of renewable energy as an enabling infrastructure for the greening of existing carbon-intensive energy – in particular the development of green steel – and of future green industries such as green hydrogen. However, the longer such promises go unrealised, the greater the risk that renewable energy development loses local legitimacy. The failure, to date, of the Pavagada solar park in providing opportunities for new industrial employment, and thus development in the region, was widely perceived as a broken promise.

Discontent can also focus on the extraction and flow of surplus away from the locality. Not surprisingly, as we document, the benefits of cheap renewable energy generally flow 'upwards' to energy retailers and large energy consumers. Retailers benefit from the falling per unit cost of renewable energy achieved for instance in the reverse auctions for power contracts. Retailers are regulated at the region-wide level and in theory are expected to pass on the benefits of falling energy costs. In practice, the gains are used to compensate for the cost of fossil fuel power generation, which remains in the energy mix, tied to existing contracts and spot market bidding systems where the price is set by the final most expensive bid. Renewables in this respect compensate for fossil fuels and keep them in the system, at least until the existing power stations close down.

Added to this, there are related problems with retailers, that are in part attributable to the growth and persistence of neoliberal policy regimes. In Germany, concerns about the power of energy retailers to manipulate energy prices have led to efforts to re-municipalise energy supply, with some success. In Australia, private retailers compete but pass on a standard cost for 'network upgrades' set by the national regulator. That cost accounted for a series of price hikes in the decade to 2013 and helped create a backlash against climate policy and renewables, especially directed at South Australia's progressive policies. In India, the state-based electricity distribution companies (Discoms/Escoms) are usually publicly owned but are heavily indebted as electricity is often not metred in rural areas and is accessed for free elsewhere, reducing the revenue base. Efforts to create a user-pays system in rural areas have had to be abandoned in a number of states, not only on technical but also political grounds, in a rare defeat for neoliberalism. In Karnataka, the advent of low-cost renewable energy appears to recoup some of these distribution losses and enables socially progressive cross-subsidies to continue in the face of market reform. Overall, these quite different institutional arrangements have generated public discontent, against private retailers in Germany, against attempts at user-pays in India, and were manipulated into a revolt against carbon pricing and renewables in Australia.

Beyond the question of energy pricing, the agglomeration of renewable energy at the region level has the broad potential to change regional identity. This may or may not be welcomed depending on the pre-existing context and on how the transition occurs. In this respect, state authorities have found new roles in redefining regional development prospects and in the process enhancing legitimacy for the transition. The key driving factor is the growing availability of excess daytime power supply from renewables at zero marginal cost, which can be stored locally or used directly to power co-located 'downstream' industries. With the growing requirement for decarbonisation beyond the energy sector, the very low cost of renewable energy can enable retrofitting across multiple sectors, from agriculture to steel production to car-making, as well as for the export trade. Regional governments increasingly seek to integrate these 'downstream' possibilities into regional development strategies geared to enhanced industrial competitiveness in a 'green' economy. Renewable energy may in this way be conceived as offering the foundation for a new model of regional and local development. Whether this will include more equitable development, beyond political rhetoric, for often disadvantaged host communities is unclear.

In the case studies we have shown how officials in energy and regional industry policy have sought to develop planning initiatives to capture the potential economic co-benefits. Such plans are not only of interest to local workers and employers – they are also (and most immediately) of interest to the renewable energy corporates seeking buyers for their electricity. With such powerful sponsors, these plans quickly gain momentum. This is reflected in the speed at which 'downstream' models are being developed and supported. One example is the effort to establish hydrogen production in South Australia, both for the local steel plant in Wyhalla and for the large lead smelter at Port Pirie, as well as for export. Similar plans are in place for coastal Karnataka and elsewhere in India, and for in-country industrial use in Germany, too. There are initiatives for large-scale storage in South Australia and Karnataka, which cater to ongoing demand for daytime power. These forms of regional industrial planning bolster confidence in renewables development, further encouraging investor interest. The model for these developments is generally centred on large-scale heavy industry rather than on more distributed possibilities, though these may gain ground over time as aggregated rooftops linked with car batteries create 'virtual' power plants able to offer round-the-clock power, as the basis for new local development strategies.

Conclusion

It has become critical for energy policy to address the wider developmental dynamics of the energy transition. It is hard to underestimate the longer-term impacts of the current transition for social change and social development. When societies transitioned from reliance on water wheels, windmills, and wood to using coal for industry, it would have been impossible to imagine what this could bring, in terms of steam trains, steamships, and the factory system, for instance, and likewise for the transition to oil and gas. Renewable energy is already on the cusp of offering virtually zero marginal-cost energy. When combined with the possibilities of distributed and scalable power, the manifold consequences are only just beginning to be explored. The last decades of renewable energy development have shown its disruptive potential: there is much more to come, not least due to the unfolding demands of decarbonisation. Policy regimes that seek to entrench the status quo will likely delay the full development of the technology's full potential and, at the same time, create edifices that in time will be far superseded.

Currently, as we document, the transition is in danger of being distorted to fit the needs of capital in the name of accumulation. Transition has been dominated by the search for a new socio-ecological fix for accumulation that can be guaranteed to generate returns over the 30-year investment horizon. The renewable energy transition entails much more than this, not least due to the emissions reduction imperative. The fossil fuel 'fix' is in the process of being superseded, and with it the structures of legitimation that it created. How that transformation happens is critical for decarbonisation and climate stability, and also for future possibilities in terms of democratic participation and social justice. The case studies we investigate in this book hint at some of these socio-political implications. We have argued that an epochal shift in transition is underway as renewables are no longer figured as an economic cost necessary for the climate, but instead are recast as an economic benefit. Cheapness gives new impetus to the sector as a competitive asset but also as the vehicle for much wider social benefit. New claims emerge seeking to socialise the potential benefits of renewable energy, against corporate capture.

Clearly, renewable energy should not displace existing livelihoods and should offer meaningful local benefits. The enormous financial surplus gained from renewable energy is extracted primarily for the power companies: local people recognise this as a great injustice and demand a change in the model. When we visited wind and solar farms, we were often told this was work in progress. A manager at Pavagada said that the plant was a 'university' in which new ideas were being tested. This 'work in progress' is now being forced in new directions. Contestation has created new demands for regulation, expanded engagement on regional development, and new initiatives in decarbonisation.

There is always the possibility of delay. Energy transition is an inevitability, but its calendar is contested. There is still an overarching dependence on fossil fuels, with substantial lock-in, not least in electricity. Strong public backing is needed to overcome these pressures: dispossession, displacement, disenfranchisement, and disenchantment can only slow the transition. Social ownership and participatory regulation remain weak yet can be the foundation for public support. As we have argued, the advent of renewables offers the possibility of both 'socialising' climate change and climatising energy. It plays a key role in advancing emerging social imaginaries in the search for climate agency, and the required ethics of climate-affected life. As with energy transitions in the past, the current juncture offers manifold (still undreamt-of) possibilities. Contestations force stronger democratic participation and point to a transition based on common ownership, beyond corporate power. The imperative is clear: social control of renewables as a public good is a foundation for climate stability, on a par with any 'common heritage' of humankind.

References

Becker, S. & Naumann, M. (2018). Energiekonflikte erkennen und nutzen (Identifying and using energy conflicts). In O. Kühne & F. Weber, eds., *Bausteine der Energiewende (Building-Blocks of the Energy Transition)*, Wiesbaden: Springer VS, pp. 509–522.

Buchner, B. et al (2023). *Global Landscape of Climate Finance 2023*, London: Climate Policy Initiative.

Bryant, G. & Webber, S. (2024). *Climate Finance: Taking a Position on Climate Futures*, New York: Agenda Publishing.

Chakrabarty, D. (2022). *A Climate of History in a Planetary Age*, Chicago: Chicago University Press.

Christophers, B. (2024). *The Price Is Wrong: Why Capitalism Won't Save the Planet*, London: Verso.

Department of Energy and Mining (2023). *Leading the Green Economy*, Adelaide: GoSA.

Department of Energy and Mining (n.d.). *Our Electricity Supply and Market*, Adelaide: GoSA.

Gabor, D. (2023). *The (European) Derisking State*, Chatlottesville: Center for Open Science.

Goodman, J. & Anderson, J. (2021). From climate change to economic change: reflections on feedback. *Globalizations*, 18(7), 1259–1270.

IRENA (2023). *World Energy Transitions Outlook 2023*, Abu Dhabi: IRENA.

Kungl, G. (2018). *Die großen Stromkonzerne und die Energiewende*, Frankfurt a.M.: Campus.

Parkinson, G. (2023). 'South Australia about to take another big step towards fossil free grid', Renew Economy, 28 March 2023.

Polanyi, K. (1944). *The Great Transformation: The Political Origins of Our Time*, Boston: Beacon Press.

Conclusions

Renewable Commons?

At a protest outside Parliament House in Australia's capital Canberra in February 2024, the former leader of the ultra-conservative National Party, Barnaby Joyce, described renewable energy as a 'swindle' and wind turbines as 'filth'. Wind farms, said Joyce in an emotional speech, 'are not farms they are factories, they are industrial dumps; environmental dumps' (Vorrath 2024b). At the same rally, the current leader of the party, David Littleproud, said that locating wind and solar farms on prime agricultural land was 'pure insanity' and suggested future renewable energy development should be limited to rooftop solar in the cities (Canales 2024). What was promoted as a national 'Rally Against Reckless Renewables' attracted only a small crowd of between 300 and 500 people; however, in 2024 senior figures in the Liberal Party, including opposition leader Peter Dutton, were seeking to capitalise on dissatisfaction with the expansion of renewable energy and transmission lines in regional Australia and the emergence of local opposition to new offshore wind projects organised by fossil fuel-funded think tanks (O'Malley 2023). Dutton is clearly targeting local anti-renewables groups and their supporters as a constituency for the conservative parties in the next federal election (Foley 2023).

In Germany, meanwhile, the chair of the neo-fascist Alternative for Germany party (AfD), Alice Weidel, has called for an immediate halt to any further development of wind power in Germany. Weidel refers to the German energy transition as 'dogma' and 'madness' and says the 'Green wind power lobby is destroying forests, nature and the environment' and 'sinning against the future of our nation and generations to come' (AfD 2023). Although the AfD's national vote has fallen below 20% in recent opinion polls after mass demonstrations against right-wing extremism, it remains the second most popular party after the Christian Democrats, and in the eastern states of Saxony, Thuringia, and Brandenburg, it leads all other parties with a current vote of over 30% (Endt et al. 2024; Tagespiegel 2024a,b). In late 2023, in the state parliament of Thuringia, the AfD joined with the Christian Democrats and Free Democrats to pass a law that will make the siting of wind turbines in forests almost impossible (MDR 2024). According to Christoph Richter of the Institute for Civil Society and Democracy in Jena, the AfD goes further than most other extreme right-wing parties in Europe in its rejection of climate protection policies and denial of climate science (Pfeifer 2023).

National political parties have not (yet) sought to instrumentalise local opposition to renewable energy in India. But elsewhere, anti-renewable rhetoric has for instance been a strong feature of Donald Trump's campaign to win the 2024 Republican nomination. Trump has called renewable energy a 'scam business', and a former senior adviser has said that 'Trump will undo everything [Joe] Biden has done' on climate policy, with the prime target being the Inflation Reduction Act (Milman and Noor 2024).

There is a clear trend for right-wing parties to exploit dissatisfaction with renewable energy and attack its social legitimacy, often with the covert support of fossil fuel companies and lobbyists (Alvarez 2022). There are also clear signs that governments are responding to these trends and the experiences of local communities that underlie them. In Australia, a Community Engagement Review commissioned by the Federal climate change and energy minister Chris Bowen and conducted by the Australian Energy Infrastructure Commissioner Andrew Dyer found there was a 'material distrust' of project developers and particularly developers of new long-distance transmission projects, caused by 'poor engagement practices experienced by landholders and community members' (Dyer 2023: 8). The report makes nine recommendations, including increasing early local collaboration, making planning and approval processes more transparent and streamlined when it comes to community feedback, and more equitable sharing of the benefits among communities (Dyer 2023: 5; Vorrath 2024a). Minister Bowen has indicated the government will accept all the recommendations in principle.

In Germany, as noted in Chapter 4, the expert commission 'Energy of the Future', which monitors the progress of the *Energiewende* and advises the Federal government, has stated that strengthening the social acceptance of wind energy installations in the countryside, in the places where renewable energy production takes place, will be decisive for achieving the necessary expansion of wind energy and, with it, Germany's climate and energy goals (Löschel et al. 2021: 25).

There is a plethora of research that identifies the sources of community opposition to renewable energy development (Betel 2020). A recent large-scale study of opposition to renewable energy projects in the US finds that 'state and local governments and renewable energy developers need to pay closer attention to the full range of socially-oriented sources of opposition to new facilities' and notes that 'the more that energy transition is delayed, the more pressing the need for utility-scale projects will become' (Susskind et al. 2022). The case studies in this book clearly demonstrate the need for these kinds of initiatives to win stronger local consent and active participation in renewable energy. To this end, governments and developers need to address the 'full range of socially-oriented sources of opposition' discussed by Susskind et al. However, we argue that more participation, democracy, and benefit-sharing of themselves will not solve the underlying problem of social legitimacy. We argue that they may ameliorate the situation but will not remove all the identified 'blockages' (Strauss et al. 2013: 10) to corporate utility-scale renewable energy. We suggest that it is the corporate character of these large-scale projects that is itself at the centre of the problem.

Corporate utility-scale renewable energy has dominated energy policy in India, Germany, and Australia, and within global institutions such as the IEA and IRENA. We argue this undermines the legitimacy of the very transition that such policies seek to promote. Corporations, by their nature, prioritise profit maximisation. This can lead to a focus on projects and strategies that are financially lucrative rather than those that align with climate objectives or with broader environmental and social goals. The interests of corporate entities, driven primarily by short-term financial gain, often clash with the sustained, long-term efforts required for effective decarbonisation. In contrast, social ownership models, involving a broader array of stakeholders, are better positioned to prioritise long-term sustainability and take-up, over immediate profits. In Europe, publicly owned electricity is found to be significantly more likely to decarbonise (Steffen, Karplus and Schmidt 2022) and beyond high-income countries, the vast bulk of renewable investment is public investment (IEA 2023). Unlike corporate entities, which may lack accountability to the public or to local communities, social ownership models that include communities, workers, and other stakeholders can introduce mechanisms for accountability and transparency in decision-making processes.

Corporate-driven projects may result in the concentration of benefits among a select few stakeholders, sharpening existing social and economic inequalities. Social ownership models aim to distribute benefits more equitably throughout the broader community, fostering community engagement and empowerment. This approach leads to more resilient and adaptive responses to challenges, as communities involved in decision-making processes are better equipped to navigate the complexities of the energy transition and address their unique needs and concerns. Corporate-driven projects may overlook environmental and social externalities associated with energy production. In contrast, social ownership models prioritise holistic assessments that consider a broader range of impacts, seeking to minimise negative consequences. Social ownership models are characterised by democratic decision-making processes and align with principles of climate justice, emphasising the need for an inclusive and equitable transition that does not disproportionately burden vulnerable communities or exploit natural resources for the benefit of a few.

The challenge, then, is not simply one of making citizens and local communities more amenable to the construction of ever-taller wind turbines and the construction of massive arrays of solar panels in the spaces and places where they live and work; in other words, smoothing out the bumps on the pathway, softening the social resistance to a neoliberal energy transition. We contend that a much more fundamental transformation is necessary, one that does not simply replicate the fantasy of unlimited energy driving continuous economic growth but seeks to understand what living within planetary boundaries might mean and how that might be achieved without simply perpetuating global inequality and the deep structural injustices entrenched in the global neoliberal order. In this reading, energy 'transitions' are not simply transitions to business-as-usual powered by sun and wind, but processes of radically reconfiguring socio-ecological relations. They offer pathways not only to the stabilisation of the Earth's climate system within planetary boundaries but also to new forms of communal life, the democratic and equitable sharing of 'nature's free gifts', and perhaps even what Jason Hickel terms 'radical abundance' (Hickel 2019, quoted in Saito 2022: 232).

Themes from the Case Studies

What would such a reconfiguration of socio-ecological relations in renewable energy involve – and which dimensions of energy transition need to be addressed? The overarching themes that emerge from the cases discussed in this book encompass spatial emptiness, the temporal evolution of renewable energy, contestation and social legitimacy, catalytic state authority, climate change articulation, environmental and social justice considerations, renewable energy neoliberalism, and renewable energy social ownership and commoning. These themes highlight the multifaceted character of the challenges and opportunities in achieving a sustainable and socially just energy future. They also prompt us to give more thought to how building this future can be based on a re-commoning of the energies of sun and wind, which would otherwise be appropriated by energy corporates under neoliberalised renewable energy. In the remainder of this final chapter, we focus again on the key themes that emerge from our study and reflect on how they might guide this agenda.

Emptiness

Renewable energy projects, such as wind and solar installations, are often situated in areas assumed to be spatially empty (waste lands) or underutilised (not fully exploited as intensively as they could be), such as rural regions or large expanses of land (or indeed ocean). This phenomenon highlights the clash between the perceived emptiness, often associated with uninhabited or underutilised spaces, and the actual presence of diverse social, economic, and ecological elements in those landscapes. These areas might be perceived as economically unproductive, having low population density, or an absence of intensive industrial enterprise, contributing to the perception of emptiness.

Controversy arises when it becomes evident that these landscapes are not empty but are, in fact, inhabited informally, offer a source of livelihood, are subject to Indigenous or other land claims, or are ecologically significant. Diverse local communities, biodiversity, and cultural or historical elements exist in these spaces. Contestation reveals the hidden layers of the landscape that are often overlooked during the initial stages of project planning. When the actual presence of communities, ecosystems, or historical sites is recognised, it can lead to local resistance against the perceived imposition of renewable energy projects. This often centres on the potential disruption of livelihoods, cultural practices, lifestyles, and social dynamics that were not initially considered by developers. Thus, energy landscapes are contested social spaces where different actors, including communities, policymakers, and corporations, have conflicting perceptions and interests regarding land use and resource utilisation. Attention to these tensions and commitment to equitable solutions is crucial for any renewable energy project.

Temporal Evolution and Learning

Each case study chapter in this book discusses the temporal evolution of renewable energy in its respective region. The trajectory involves phases of development, including the establishment of small-scale projects, transitions to larger-scale projects, and, in some cases, the emergence of hybrid installations combining different renewable sources. The conclusions of each chapter reflect on the challenges faced during the renewable energy transition and identify lessons learned from the experiences in the respective regions. This includes the need to address structural inequalities, ensure equitable access to benefits, and learn from past phases of renewable energy development. These all suggest the need for a nuanced and context-specific understanding of the renewable energy transition. They underline the complexity of the issues involved and the importance of considering historical, economic, and social dynamics in crafting effective policies and strategies for energy transition.

Contestation and Legitimacy

While not every renewable energy project we studied involved contestation, it is a shared theme in all case study chapters. The chapters delve into how these projects are forces of change within the socio-economic fabric, creating differential opportunities, detriments, and challenges for various stakeholders in host communities. Attachment to land and landscape, distributional conflicts, access to benefits, and the role of local governance are explored as critical aspects. The ethnographic analyses highlight the social legitimacy of the energy transition as a critical factor, with different dimensions of contestation influencing the social acceptance and success of renewable projects. In host communities, the importance of renewable energy and the need to address climate change via emissions reduction are rarely questioned. The social legitimacy of the new industry depends on the particular ways in which energy companies and state regulators are responsive to the stresses and strains imposed by project development, particularly in already disadvantaged host communities.

Catalytic State Authority

The role of the state in regulating and governing renewable energy is discussed in all three chapters. We examine state intervention in each case, with a focus on how national and sub-national governmental agencies act as arbiters, facilitators, or promoters of renewable energy, with consequences for the overall trajectory and legitimacy of the transition. In the German case, it is within the regional planning bodies that the key decisions are made, and contestations take place. Indeed, we argue that it is primarily through these regional planning bodies that the energy transition in Brandenburg is regulated and governed; they are the principal arena for the 'social making of arguments', and with it, the social contest over legitimacy. In India we find quasi-state agencies taking the lead, meditating fields of public authority and private power, and convening agendas for renewable energy development. These agencies play a central role in enabling social consent for projects, gearing their initiatives to anticipate public concern. In this, public initiative can play a critically disruptive role. We observed this in the early phase of roof-top solar panels in South Australia, which overwhelmed expectations and is now more highly regulated by the national Energy Market Operator. The 'theft' of electricity in India is another example, where citizens met their energy needs in defiance of state oversight.

Climate Change Articulation

Climate change propels energy transitions via national and international policies and agreements. At the sub-national level, these commitments are often a background factor, taken for granted, and generally not disputed. We show how that national context sets and frames sub-national agendas, suggesting its centrality is structurally articulated into the process of energy transition. The extent to which climate change is mobilised as a reason for advancing energy transitions varies across the three cases. In Germany, at least until now, the Energiewende has been a central part of the national narrative of climate action, with generally broad support from the majority of citizens (Lafrance and Wehrmann 2023). A national survey in Germany in 2023 showed that 92% of respondents support a faster expansion of renewable energy (Amelang 2023). In India, under Prime Minister Narendra Modi, the energy transition has become part of national development discourse. Recent national surveys show a majority of respondents see climate change as a threat, although there are differences according to location and level of education (Vigers 2023). In Australia, the 'climate wars' stalled climate policy until 2022 with the election of a Federal Labor Government; until then, the energy transition unfolded primarily through region-level action by state governments (and most especially by South Australia). Citizens have long been ahead of denialist governments, with a consistent majority of respondents to annual surveys stating climate change is a 'serious and pressing problem' that requires 'taking steps now even if this involves significant costs' (Lowy Institute 2023).

While citizens in these and many other countries are willing to modify behaviour at an individual and household level where feasible, international surveys show they are critical of governments that are not sufficiently proactive on climate action, especially high-income countries (Narawad 2023). In our three case study regions, the focus is on safeguarding livelihood and environment as well as maximising benefits when large-scale renewable energy projects arrive. These projects are no different from any other large construction or infrastructure project in this respect. Moreover, communities may be dealing with various challenges, such as economic downturns, social disadvantage, poor services, or public health issues that overshadow the longer-term implications of climate change and lead residents to interpret renewable energy as imposing a cost on their locality as a renewable energy 'sacrifice zone'. In response, there is a need to more deeply socialise climate, allowing climate change to become more strongly articulated with everyday concerns and, as such, foster more meaningful action at the ground level.

Environmental and Social Justice

The chapters highlight the need for a broader transformation beyond decarbonisation. We highlight participants' concerns related to environmental and social justice, emphasising that the imperative of decarbonisation involves a structural transformation that must address inequality and disempowerment in many social domains. The chapters explore how public involvement in decision-making and in ownership structures is critical to achieving the social legitimacy of the transition.

As we reflect on the insights gleaned from the ethnographic chapters in Brandenburg, Karnataka, and South Australia, the concept of renewable energy regionalism emerges as an important analytical tool. The themes emerging from the region-level narratives of renewable energy transcend local specificities and unveil shared challenges and opportunities in the pursuit of a sustainable and socially just energy future. The spatial assumption of emptiness, a recurring motif, underscores the clash between preconceived notions of underutilised or underexploited landscapes and long-established social, economic, and ecological attachments and interests.

Each of these five themes emerging from the case studies – emptiness, temporal evolution, the centrality of the state, climate change articulation, and environmental and social justice – focuses our attention on the inequities and contradictions that define the investment in large-scale solar and wind farms. Exposing these helps reveal and foreshadow potential contestation hotspot issues, allowing more reflexive renewable energy policymaking.

Renewable Energy Neoliberalism

Support for the development of large-scale corporate-owned solar and wind farms is premised on transforming ecological value into energy capital. This amounts to an appropriation of the commons to subsidise financial accumulation. With social relations reordered and ecology transformed, the process closely mirrors Marx's concept of primitive accumulation: it literally smashes apart previously existing socio-ecological relations, in this case coal-fired power stations, in favour of the new model (Saito 2022). The ecological 'frontier' of accumulation is dramatically extended to subsume wind and solar power into the process of accumulation. To achieve this, pre-existing social relations to land and ecology are swept aside; this, as Harvey highlights, is commonly experienced as a process of 'accumulation by dispossession' (Harvey 2004). In each of the case studies, the allocation of resources to facilitate the flow of energy – and capital – is premised on neo-liberal conceptions of land, ecology, and labour as commodities to be exchanged through market transactions. Here the state identifies land that is necessary for the installation of solar panel arrays and wind turbine clusters, as well as transmission lines and distribution connections.

With the help of state authorities, renewable investors may gain access to the free resources of solar and wind power, but they can quickly become victims of their own success. The abundance of daytime solar and wind energy can vastly exceed energy demand, deflating prices. Prices can fall to zero and even become negative. As investors are unable to gain an income for their product, the sector faces a 'realisation' problem. As discussed in the introduction to this book, IRENA now cites this as a major and growing dilemma for privately financed renewables. Attempts to restore investor returns centre on 'curtailing' supply. In Australia for instance the Energy Market Operator has directed retailers to block households with rooftop solar panels from feeding electricity into the grid when large-scale solar and wind farms are generating more power than can be sold. Literally wasting renewable energy in this way clearly negates the process of energy transition yet is deemed a necessity to maintain investor 'confidence'. A system that depends on private investors cannot allow zero or negative pricing.

Indeed, across the renewable energy transitions in Brandenburg, Karnataka, and South Australia, there is an overarching narrative of neoliberalisation. As the regions grapple with the imperative to decarbonise and embrace renewables, the neoliberal character of their respective transitions pre-ordains the structure of the energy landscape. The emphasis on capitalising ecology and transforming nature's 'free gifts' into a bankable asset echoes neoliberal principles of commodification and privatisation. The process of encasing ecology as a resource, with corporations holding exclusive rights for exploitation, mirrors neoliberal tenets of economic efficiency and individual enterprise.

In the process, energy transition is captured for corporate interests and not for public benefit, nor even for climate stability. We have highlighted how the sector's competitiveness comes into direct conflict with its purpose as a vehicle for wider social benefit, including in terms of meeting the need for abundant energy in the process of society-wide decarbonisation. The neoliberal construction of the renewable energy transition, evident across the diverse socio-political settings, lays bare the inherent contradictions between private accumulation and the pursuit of socio-ecological justice and climate stability. We have tracked how this collision is manifested on the ground in terms of contestation and re-regulation, and how this points to new forms of social ownership and participation.

Renewable Energy Commoning

Across the case studies, we see not just the imperative for energy commons but its unfolding logic in present-day transitions, as an immanent trend. The case studies point to challenges and alternatives to neoliberal transition, from recent policy initiatives for cooperatively owned solar farms in Brandenburg, to Indigenous Traditional Owners taking equity in a large-scale solar in South Australia, to local landowners in India demanding development gain from wind farms and shared equity in large-scale solar. These outcomes are becoming increasingly prevalent as local populations become more organised and, in response, governments and private investors take new steps to respond. This development is not unique to the cases we examined: a recent study documents different models of indigenous and local community partnership in renewable energy in Canada, Chile, Colombia, Honduras, Mexico, Norway, Scotland, Sweden, and the US, finding a series of parallel developments (IHRB 2023).

As we have argued, there are many opportunities opening up in the unfolding transition. The advent of free or nearly free daytime electricity now offers the foundation for a genuine 'energy commons', and for a much wider energy transformation (Lifshitz-Goldberg 2010). Constructed scarcity – though state-imposed curtailment – negates the possibility of this transformation (see Hickel 2019). As renewable energy becomes more predominant, this contradiction is only likely to sharpen. Indeed, with rapid technological development in the field, we can expect a wider embrace of more distributed forms of renewable energy, enhancing abundance. We can expect this will radically disrupt and overwhelm the current model of the state-backed large-scale solar and wind farms and open more scope for household and community-based energy generation and distribution systems, further underpinning the vitality of the commons.

Across the board, there is a generative and escalatory momentum away from renewable energy neoliberalisation and towards renewable energy commoning. More equitable distribution of benefits advances both social justice and sustainability. Social ownership provides a framework that places decision-making power in the hands of diverse stakeholders, fostering a more inclusive and equitable transition, allowing alignment with sustainability, social justice, and genuine decarbonisation goals. Heightened social ownership concurrently underpins democratic participation in the emerging transformations, fostering a deeper sense of community stewardship, nurturing sustainable practices, and commitment to the long-term viability of renewable energy assets.

Recommendations

On this basis we can seek to advance normative recommendations enabling the renewables commoning agenda, emphasising the crucial role of ownership structures in the energy transition.

First, social control of renewable energy offers a set of conditioning factors beyond the legitimacy deficits of neoliberalised renewables. The central goal of sustainable, just, and inclusive energy transitions should be to create utility-scale renewable energy that is publicly owned, whether national or municipal-owned, community-owned or cooperative. These should not only create local jobs and promote local economic development but also provide a reliable source of livelihood (IHRB 2023). Public involvement and democratic decision-making lie at the core of this vision. Embracing a newly proactive role for state authorities capable of enabling socialisation and rapid expansion of decarbonisation is essential.

Second, the priority of environmental stewardship, in particular emissions reduction, should be paramount in any model for renewable energy. The aim should be to encompass energy transformation across social life rather than simply replace one source of fuel by another. This amounts to an invitation to rethink the place of 'nature's free gifts' as a foundation for the remaking of the socio-ecological order. Knowledge sharing and education may play a crucial role in this, in terms of fostering social engagement with renewable energy projects.

Third, structures of collaboration need to be strengthened to promote shared responsibility and mutual support across society in decarbonising energy systems. This can underpin social bonds, for instance, via collective involvement in publicly run utility-scale renewable energy projects. Inclusivity and diversity can be championed within publicly owned and run structures, ensuring that decision-making bodies and project participation reflect the richness and diversity of society. This approach can foster a more inclusive and representative approach to renewable energy development.

Fourth, the scalability and replicability inherent in renewable energy should be enhanced for multilevel renewable energy transformations. We would argue for a vision of utility-scale renewables that complements rather than competes with other scales of 'renewability' (Haney and Pollitt 2013). Distributed renewable abundance is needed at all levels, from hand-held devices, to household appliances, to community, municipal, and utility scales. We would seek a cumulative socially generative energy transformation that aligns with principles of community empowerment, equitable benefits, local ownership, and environmental stewardship, ultimately contributing to a more sustainable, just, and inclusive energy future.

Fifth and finally, we would seek closer alignment with the critical socio-ecological discourse that challenges the notion of 'green growth' and emphasises the importance of societal structures that enable reduced and more efficient use of energy as central to achieving decarbonisation. Again, as discussed in this book's introduction, IRENA recognises that reduced energy use is a critical component of energy decarbonisation. This perspective challenges the conventional wisdom that economic growth is synonymous with progress and sustainability; instead it points to the need for a re-evaluation of our economic and energy systems, emphasising local empowerment, ecological responsibility, and a more holistic understanding of well-being.

Degrowth is an important corrective to the (still) powerful assumptions of ecological modernisation. Large-scale privately owned renewables without wider social change that can address energy use as a 'social commons' is unlikely to deliver (Bryne, Martinez and Ruggero 2009). As Kohei Saito has argued, fantasies of endless growth underlie not only the 'ecomodernist' faith in the innovative power of neoliberal capital to overcome the climate emergency but also the

Promethean programs of twentieth-century 'productivist socialism' and 'left accelerationism'. Saito notes growth assumptions now 'sneaking into' recent attempts to renew the 'vision of a post-capitalist future' (Saito 2022: 232). To these fantasies, Saito counterposes a future grounded in 'degrowth communism' – one which seeks to stay within planetary boundaries while satisfying basic human needs and seeking to scale down the economy to create a more sustainable and egalitarian society (Saito 2022: 242).

The five broad proposals advanced here grow out of the insights that have emerged from our ethnographic case studies. As we have shown, in each of the three regions, a dialectical process of social action and learning has emerged in the process of contesting large-scale corporate renewables. Energy has been politicised in new ways and new claims have been advanced, forcing reflexive re-regulation across the sector, along with new forms of participatory engagement and more proactive regional renewables planning. Project developers and governments have responded to claims, leading to experiments with new forms of benefit-sharing, ownership, and energy governance. We understand these as generating a range of agendas for 'non-reformist reform', opening up the possibility of transformations for energy commoning, in the sense advocated for by André Gorz (1967).

The final de-growth proposal is an extrapolation, inferring from the ethnographic research (see Abram et al. 2023). The lessons learnt from our case studies point to a horizon beyond the present: they suggest, in our view, that the transition to renewable energy opens a wider and deeper perspective of societal transformation, pointing to a much larger transition beyond capitalist growth. Energy transition is highly generative, and the evolution of socio-ecological relations in each case is dynamic and ongoing. There are, in particular, clear pressures beyond renewable neoliberalism. These, we argue, point to a process of renewable commoning. We anticipate a process of engaging with the growth contradictions of decarbonisation, and with the rich field of discussion around ideas of degrowth, from Saito's 'de-growth communism' to IRENA's model of reduced energy use and declining growth-energy intensity. The continued failure to reduce greenhouse gas emissions creates an imperative to address these issues of growth and emissions, as 'an emerging scientific research program and a political project' (Schmelzer, Vetter and Vansintjan 2022: 297).

Here we return full circle to the Asimov story with which we began this book. *The Last Question* is not only a fantasy of 'all the energy we can possibly ever use for free … forever and forever and forever' (1956: 8) – it is also a fantasy of endless and unlimited growth. The seemingly limitless energy of the sun, 'stored, converted, and utilised directly on a planet-wide scale' allows humankind to leave planetary boundaries behind and colonise the solar system, the galaxy, and then the entire universe. Implicit in the technocratic fantasy of transitioning to a world powered by renewable energy is an assumption that growth and 'business-as-usual' will simply continue in a decarbonised global economy. The fantasy may seduce some, but increasingly climate change is forcing a reckoning, requiring society to address this, our own 'last question'.

References

Abram, S., Waltorp, K., Ortar, N. & Pink, S. (2023). *Energy Futures: Anthropocene Challenges, Emerging Technologies and Everyday Life*, Berlin: De Gruyter.

AfD (2023). 'Alice Weidel: Grüne Windkraft-Lobby zerstört Wald, Natur und Umwelt – und die CDU macht mit', Alternative fur Deutschland, Media Release, 6 December 2023.

Alvarez, G. (2022). 'Fossil-Fuel Funded Opposition Is Blocking America's Clean Energy Transition', Forbes, 30 November 2022.

Amelang, S. (2023). "Overwhelming majority" supports acceleration of renewables roll-out in Germany – survey', Clean Energy Wire, 5 April.

Asimov, A. (1956). The Last Question. *Science Fiction Quarterly*, 4(5), 6–15.

Betel, S. (2020). Research on the social acceptance of renewable energy technologies: past, present and future. *Energy Research and Social Science*, 68, 101544.

Bryne, J., Martinez, C. & Ruggero, C. (2009). Relocating energy in the social commons. *Bulletin of Science, Technology & Society*, 29(2), 81–94.

Canales, S. (2024). 'Current renewables rollout "pure insanity", Littleproud says', The Guardian, 6 February 2024.

Dyer, A. (2023). *Community Engagement Review Report, Australian Energy Infrastructure Commissioner*, Canberra: Department of Climate Change, Energy, the Environment and Water.

Endt, C. et al. (2024). 'Warum die AfD plötzlich schwächelt', Zeit Online, 2 February 2024.

Foley, M. (2023). 'Farmers' lobby swings behind Dutton on campaign against wind farms', Sydney Morning Herald, 31 October 2023.

Gorz, A. (1967). *Strategy for Labor. A Radical Proposal*. Translated, M. Nicolaus & V. Ortriz, Boston: Beacon Press.

Haney, A. & Pollitt, M. (2013). New models of public ownership in energy, *International Review of Applied Economics*, 27(2), 174–192.

Harvey, D. (2004). The 'new' imperialism: accumulation by dispossession. *Socialist Register*, 40, 63–87.

Hickel, J. (2019). Degrowth: a theory of radical abundance. *Real-World Economics Review*, 87, 54–68.

IEA (2023). *Scaling up Private Finance or Clean Energy in Emerging and Developing Economies*, Paris: International Energy Agency.

IHRB (2023). 'Community ownership of renewable energy: how it works in nine countries', Just Transitions Report, Institute for Human Rights and Business, 21 February 2023.

Lafrance, C. & Wehrmann, B. (2023). 'A shared challenge: attitudes about climate action in France and Germany', Clean Energy Wire, 25 July 2023.

Lifshitz-Goldberg, Y. (2010). Gone with the wind? The potential tragedy of the common wind. *Journal of Environmental Law and Policy*, 28(2), 435–470.

Löschel, A., Grimm, V., Lenz, B. & Staiß, F. (2021). *Stellungnahme zum achten Monitoring-Bericht der Bundesregierung für die Berichtsjahre 2018 und 2019*, Berlin: Expert Commission for the Energy of the Future Monitoring Process.

Lowy Institute (2023). *Lowy Institute Poll 2023*, Sydney: Lowy Institute.
MDR (2024). 'After a week-long examination: The disputed Thuringian Forest Law can come into force', Mitteldeutsche Rundfunk (MDR), 7 February 2024.
Milman, O. & Noor, D. (2024). '"In a word, horrific": Trump's extreme anti-environment blueprint', Guardian, 6 February 2024.
Narawad, A. (2023). 'Global surveys show people's growing concern about climate change', Clean Energy Wire, 27 November 2023.
O'Malley, N. (2023). 'Projects hijacked by "political actors like Peter Dutton suddenly pretending to care about whales"', Sydney Morning Herald, 4 November 2023.
Pfeifer, H. (2023). 'Rechter Höhenflug in Deutschland – die AfD und das Klima', IM FOCUS, 31 May 2023.
Saito, K. (2022). *Marx in the Anthropocene: Towards the Idea of Degrowth Communism*, Cambridge: Cambridge University Press.
Schmelzer, M., Vetter, A. & Vansintjan, A. (2022). *The Future Is Degrowth: A Guide to a World beyond Capitalism*. London: Verso Books.
Steffen, B., Karplus, V. & Schmidt, T. (2022). State ownership and technology adoption: the case of electric utilities and renewable energy. *Research Policy*, 51(6), 104534.
Strauss, S., Rupp, S. & Love, T. (2013). *Cultures of Energy: Power, Practices, Technologie*, London: Routledge.
Susskind, L., Chun, J., Gant, A., Hodgkins, C., Cohen, K. & Lohmar, S. (2022). Sources of opposition to renewable energy projects in the United States. *Energy Policy*, 165, 112922.
Tagespiegel (2024a). Neue Wahlumfrage: AfD in Brandenburg, Thüringen und Sachsen weit vorn, Tagespiegel, 11 January 2024.
Tagespiegel (2024b). Niedrigster Stand seit Juli: AfD rutscht in Umfrage auf 20 Prozent, Tagespiegel, 4 February 2024.
Vigers, B. (2023). 'Is India prepared to change with its climate?' Gallup, 5 September 2023.
Vorrath, S. (2024a). 'Rally betrays anti-renewables desperation, as Joyce calls on "army" to go home and recruit', Renew Economy, 6 February 2024.
Vorrath, S. (2024b). 'Call for developer rating scheme, as review finds "material distrust" of renewables, transmission', Renew Economy, 2 February 2024.

Index

www.ingramcontent.com/pod-product-compliance
Lightning Source LLC
LaVergne TN
LVHW062209200926
848484LV00006B/1355
* 9 7 8 1 0 0 9 4 8 5 6 1 6 *